BUTLER AREA PUBLIC LIBRARY
BUTLER, PA.

PRESENTED IN MEMORY OF

Carl H. McCall

by

A.K. Steel Melt Shop Maintenance

Wait 'Til the Cows Come Home

Richard Triumpho

Sunnyside Press
St. Johnsville, New York   Berkeley, California

WAIT 'TIL the COWS COME HOME
Farm Country Rambles With A New York Dairyman

Copyright © 2002 by Richard Triumpho
All rights reserved

The text of this book, in different form, appeared originally in Hoard's Dairyman magazine and is reproduced here with the kind permission of the editors at W.D. Hoard & Sons

ISBN 0-9717214-0-8

Publisher's Cataloging-in-Publication
(Provided by Quality Books, Inc.)

Triumpho, Richard.
 Wait 'til the cows come home / Richard Triumpho. --
1st ed.
 p. cm.
 Selections from author's contributions to Hoard's dairyman magazine

 1. Triumpho, Richard. 2. Farmers--New York (State)--St. Johnsville--Biography. 3. Dairy farming--New York (State)--St. Johnsville. 4. Farm life. I. Title.
II. Title: Hoard's dairyman.

SF229.7.T7A34 2002   636.2'142'092
             QBI02-200167

Printed and bound in the USA

SUNNYSIDE PRESS
St. Johnsville, NY   Berkeley, CA

CONTENTS

SPRING

1	Bessie Ferguson	9
2	Pollywogs	15
3	Beaversprite	21
4	Calendar	28
5	Those Darn Goats	32
6	Sheep Power	38
7	Hardwoods in May	42
8	Horse Training	46
9	Gus	50
10	Cow in the Well	54
11	Something of Interest	58
12	Jefferson Dodge	62
13	Oh, Beans!	68
14	Maple Bees	72

SUMMER

15	Fred's Ultra-Light	79
16	Tiffy's Foal	83
17	June Haying	87
18	Chlorine Spill	91
19	Fred Goomsby	95
20	Cattle Crossing	100
21	Hayfield Bird Song	104
22	The Campers	108
23	Rural Cemetery Census	112
24	Summer Vacations	118
25	That Darn Bull	123
26	Rooftop Raspberries	128

AUTUMN

28	Power Line R.O.W.	141
29	Bill Keck	146
30	Barns Into Homes	150
31	Sam the Cow Mat Man	154
32	The Cattle Dealer	158
33	The Old Jeep Truck	163
34	Killdeer	169
35	Three Dollar Bill	173
36	Corn	177
37	Farm Art	182
38	Russian Truck Jumping	186
39	Dynamite	190

WINTER

40	Ray Nagele	197
41	The Deer Tracker	204
42	Frozen Pipes	209
43	Two Doors From George Bush	214
44	Loren and Nellie	218
45	Swampy Walrath	223
46	Competition	228
47	Tiny Shuster	232
48	Coyote	238

FIELDS AFAR

49	Southern Cross	245
50	Spring Terrace farm	249
51	Wales	255
52	Loch Lomond	261
53	Kings Park Farm	267
54	Mongolian Adventure	273
55	Alaska Bush Pilot	278
56	Point MacKenzie	283

Preface

Thirty years ago, in the autumn of 1973, I worked up the courage to send a query letter to the managing editor of Hoard's Dairyman, Eugene Meyer, suggesting that while his magazine did a fine job of addressing the business side of dairying, it neglected the pleasurable side — after all, many farm families consider the benefits of a rural lifestyle a major reason for putting up with the hard work of farming. I suggested that a new column, written by myself, celebrating the pleasurable moments of dairy farming (as well as the rueful ones) would strike a chord with readers and be a valuable addition to his magazine.

He kindly wrote back that there didn't seem to be room for a new column; they already had a Hog Column, a Veterinary Column, and so forth. Undaunted, I sent him a few pages of my writing, in the style of a day-by-day journal. To my everlasting amazement and gratitude he liked those samples; thus began my "Jottings in a Dairyman's Journal" and three decades of pleasurable association with Hoard's Dairyman.

The "Dairyman's Journal" became so popular that, in response to readers' requests, the editors assembled a selection of my columns from 1973 to 1979 and published them as my first book, *No Richer Gift*. This new volume, *Wait 'Til the Cows Come Home* is a further collection of selected columns from 1980 to 1992

It was through my "Jottings" that I became acquainted with dairy families all over the world. My first trip to New Zealand was from an invitation by Gerald and Annette Gunther to their Spring Terrace Jersey farm on the southern tip of the South Island, near Invercargill. That was twenty years ago; our daughter was eleven, and PanAm had a special two-for-one rate where children eleven and under could travel practically free —an offer we couldn't refuse. The Gunthers' management intensive grazing impressed me so much that on my return home I began converting my old barbed wire pastures to a hi-tensile-electric paddock system in 1982.

That trip made me a confirmed traveler, and no matter where I went worldwide I found Hoard's Dairyman magazine respected as 'the bible of the dairy industry.' Every summer's brief vacation led to new far-away places: motoring through the Swiss Alps to visit mountain farmers; touring England, Scotland and Wales, staying in Bed-and-Breakfast farm homes; driving the beautiful Alaska Highway to see the new dairy farm acreage the State of Alaska was carving out of the wilderness at Point MacKenzie near Anchorage.

Next came a ten-day farm tour of the Soviet Union, sponsored by US Exchanges. A year later I returned through the back door of Siberia, across the Bering Strait. A bonus for me was finding a dairy farm in the bleak tundra landscape near Provideniya. The following year, after the breakup of the Soviet Union, found me again in Siberia, backpacking in the mountains of the Buriyat Republic southwest of Lake Baikal; then touring Mongolia by Russian Jeep for a week, visiting nomadic Mongolian farmers with their herds of cattle, horses and sheep, staying in yurts, sharing their meals, and coming home to write about these adventures.

SPRING

BESSIE FERGUSON

The March wind roars, blasting another snow squall at my truck as we turn off the blacktop highway to the narrow dirt road through the woods, past the sign painted in bold black letters, LASSELSVILLE STATE FOREST 1100 ACRES. Next to it is a rusted road sign, barely legible: SCHULLENBURG ROAD. On my pickup are enough bales of hay for Bessie Ferguson to carry her sheep through the remaining winter weeks. Bessie had stopped at my barn a few days ago and asked if I could spare some hay; her own supply was running low and her ewes were due to lamb soon. So this morning as soon as I finished milking the cows, Derik, my hired man, helped me load bales; he was riding along to help unload them in her barn.

Tall forest of second-growth hardwoods intermingled with planted conifers crowds my truck on either side. Whenever I happen to drive this road it's hard for me to believe this woods was all farmland a century ago, but the evidence remains: Beneath the bare trees are old stonewalls running like molehills under the snow. I think of the back-breaking labor that went into cutting the virgin forest, then clearing the land of those stones — many of them massive boulders — and laying the stonewalls to mark the fields.

The Scotch-Irish and German immigrants who settled here farmed for only a hundred years before the land petered out; by the early 1900's most of the farms were abandoned. During the Depression years of the 1930's the Civilian Conservation Corps planted spruce and pine seedlings to help the hardwoods reclaim the farm land; now this mature forest makes it seem those farms had never been..

The town snowplow has been up through here already early this morning. The town has to keep the road plowed although there are only two remaining residents — Bessie, and her neighbors, the Maddoxes.

Twisting and curving through the woods, and all the while rising uphill, the snowplowed path unwinds before the truck for two miles, then abruptly ends where the plow has turned around at the Maddox place just ahead. I slow down and shift into four-wheel drive.

Bessie's farm is two hundred yards down the unplowed lane to the right. Now eighty-two years old, a widow for thirty years, and fiercely independent, she lives alone back here in the woods on the place where she was born. Although no electric or phone lines ever came to the early settlers in this remote area, her farm in its prime days had thirty cows.

After Bessie was widowed she kept the farm going, doing most of the rugged work herself; a hired man helped with the summer haying. Eventually the cows were sold and now her income is from a small flock of sheep. Her only other livestock consists of some chickens, a milk goat, and three dogs. It must be sad for her, with the dairy cows long gone, to see the hayfields growing up in brush and the forest slowly closing in.

As I back the truck up to the small sheep barn, Bessie comes out of the house, leaning on a hickory stick for a cane. This cold, windy weather aggravates an old back injury of hers that she got cutting firewood. She is wearing a sheepskin cap with ear-lappers, a heavy wool jacket, blue jeans tucked into brown rubber boots. Derik and I carry the hay bales in the barn and stack them for her. When we are done, Bessie says "Come in the house and sit for a spell and warm up."

There is a wood stove in the kitchen, a kerosene lamp on the table. On the wall above the kitchen table is an old frame with a faded photograph of a bearded man. On the opposite wall is a similar old daguerreotype portrait. I point to the picture over the kitchen table and ask who it is.

"That's my grandfather on my mother's side, John C. Shullenburg."

I point to the other portrait and ask who he is.

"My grandfather on my father's side, John Shullenburg."

I look perplexed until she explains.

"My father and my mother were cousins. Both named Shullenburg. There was four John Shullenburgs here: John C.

Shullenburg, John Shullenburg, John Henry Shullenburg, and Long John Shullenburg.

"Didn't you have an uncle or some relative who was a veterinarian?"

"My father's father! Studied for a veterinary. Had everything ready to go to work, and threw it up!

"When they first come over in here, they come over from Germany, and my father's father got here first. He landed in the village down below, and them there Sanders that used to be down there, they owned land up here, my God! Oh! I guess so! This whole darn territory!

"Anyway, my grandfather got in with them down there, and here's where they put him — right up here on the hill to start cuttin' timber. So he went to work for them. And as soon as he got any cut off, they give him the ground. So he just kept right on — goin', goin', goin'. That's how he got started ownin' land in here. Worked night and day. Raised a girt big family. Wasn't but fifteen of them. Course, there was quite a few of them died when they was young."

"What kind of a farm did they make here?" I ask. "Did your grandfather have cows too?"

"Yaass! Mercy yes! Carried the milk to the cheese factory. 'Twas over there. Lotville I guess it was. My grandfather up here on the hill, he had a whole drove o' cattle and a whole lot o' sheep — a girt big flock o' sheep — all that stuff, right up here. This was a big place here, but now it's all growed up to brush. All died out, got down to nothin'."

"What kind of cows did they have?"

"Regular Holsteins and Jerseys, and some o' them little black ones — what they called black Jerseys. Good milkers they was."

Spring

Bessie goes to the stove, lifts the lid, pokes the fire with the lid handle, and puts another chunk of wood in. The ash door pops open and a big red, glowing ember almost falls out.

"Whoa here!" She pokes it back.

I ask her if this whole area was farms back then.

"This here was a big farm, and this is where my father and mother was; my father's father up on the hill there; and the Williamses down below. Yaass, big place!"

"Did everybody have cows?"

"Yaass, cows, cows, cows!" She laughs. "If they had only a little bit of a spot, they'd have a couple o' cows. Some of the settlers didn't have much of a farm—they'd have one or two cows and a few sheep. Chickens 'n' geese and all sich stuff. Messin' with feathers and the little wool they had—most o' them spun that themselves—they used it. My grandmother up here on the hill, she spun yarn, made their socks and mittens an' everythin'. I don't know how they got all that work done."

"How come everybody stopped farming and moved away?"

"Ho! City! Got city crazy! God almighty, I couldn't see that. They called me a fool and everything because I stayed right here. I can't see the city—I hate that place! Boy, I can't stand it there—just long enough to get my groceries 'n' stuff and get out! That's as long as I can stand it. I don't want anything to do with it!"

Bessie shifts the chair closer to the kitchen table and leans her arms on the worn oilcloth. I ask what year she was born, and her laughing eyes, young and sparkling with life, light up her wrinkled old face.

"1907. The sixth of June. A June bug." She chuckles. "I was just a little girl when that Williams bought that place over there, west o' the road. He, his wife, and two sons. Charles Orlando Williams. His initials was COW!" She laughs at the memory.

"Yeah. They called him Cow, and they called him Old Goose Egg, 'n, oh, my God! What names! He farmed in there some."

Again the kitchen is quiet. A chunk of wood in the stove hisses, sighs and settles.

"Well, when Orlando's wife passed away, he sold her body. Got five hundred dollars for her dead body."

Astonished, I raise my eyebrows. " How awful! That's incredible!"

"And then when his sons died, he sold them too. Boy — everybody talked about that round here — they thought that was the worst thing ever was!"

"Sold them too?" Aghast, I inquire, "For what?"

Bessie smiles. She is enjoying the effect the story has on me. "Well — then he had the money to spend!"

"I mean, who would buy them? For what?"

"Them — that there — where doctors study; whatever they call that thing."

"Medical school?"

"Yeah. And they take everything like that they can get."

I shake my head and make a wry face.. "What kind of a man would do a thing like that? To his own wife and sons!"

The kitchen is quiet; wind rattles the windowpane. I can hardly comprehend this macabre tale, but Bessie isn't finished.

"So, when Orlando died, they comes and gets his body too."

This is too much. "Huh? What? I'll be! Who got the money for that?"

Bessie can barely contain her laughter. "Oh, he had that spent before he died!"

POLLYWOGS

Spring comes in with high winds and high hopes, with days of brilliant sunshine and white clouds racing their own blue shadows across the verdant hills of our farm, and cows lying in green pastures, their jaws moving rhythmically in bovine contentment while overhead the winds do shake the darling buds of maple trees. There's a feeling of freedom now for both a dairyman and his herd to be outside in the fields after a long winter in the barn.

Today I finished the last of the fence mending, undoing the damage wrought by five long months of winter. I tightened sagging wires that had been loosened by the weight of snowdrifts. Fenceposts heaved up by the frost I straightened and drove in deeper with the post maul. Rocks that had tumbled from the stonewall I lifted and set right again. It was a joy to work with the west wind full of the scents of spring: the smell of warm earth and green grass; a joy to hear the bird songs carried on the wind — of redwing black birds and song sparrows returning to their nesting in the meadows. I remembered to look in the brown leaf mold behind a certain fencepost where the wild asparagus, urgent in spring, always thrusts its spears; I plucked them for supper.

I finished fence mending in late afternoon, in time to fetch the cows from pasture for the evening milking. The herd ambled up the lane toward the barn and stopped at the barnyard water trough. They lowered their heads to drink but then I saw them hesitate and stare at the water , so I went over to see what they were looking at.

Beneath the surface there was a flash of orange, a rippling of fins. Some circumstantial evidence is very strong, as when you find a goldfish in the water trough!

A school of tiny minnows followed the goldfish. In their wake swam a covey of pollywogs. It was spring all right! I glanced from the water trough to the marsh down in the creek pasture. Sure enough, two small figures were crouched low over the water: Zachary, the seven-year-old neighbor boy, was helping my nine-year-old daughter, Ann, catch more pollywogs. When they had their coffee cans full, they carried them up from the pasture.

The barnyard water trough was their giant aquarium. It was more exciting than the glass fish tank Zachary had at home. He got the idea for bringing the goldfish when he and his parents went away for spring vacation. The goldfish would have starved in the home fishbowl while the family was gone for two weeks. What could be more practical than to bring the goldfish to the water trough where Ann could feed them every day.

When Zachary returned from spring vacation he decided to leave the goldfish in the water trough. And there were more pollywogs for the two children to catch. They ended up with three varieties of these tadpoles, all of them looking like miniature whales. The most abundant were those of the ordinary green frog; they were about the size of a kernel of corn. The marsh water swarmed with them.

One day in a quiet pool the two kids found some gigantic tadpoles; they were twice as large as a kidney bean. Obviously they were bullfrog tadpoles. They were not as numerous, so the kids brought only half a dozen to the water trough.

Ann and Zachary were excited the day they found the miniature tadpoles; they scooped them in their coffee cans and dumped them in the water trough. Ann called me from my work: "Dad, come and look at what we found!"

The little critters were tiny, smaller than grains of rice. "Those must be pollywogs from the spring peepers," I said. "Do you remember the night last month when we went to the swamp and saw the peepers?"

It was an April evening; supper over, we were sitting on the porch after dark, listening as the spring peepers began tuning up their symphony. The swamp pond was two hundred yards from the house, but their music drifted to us clearly on the damp night air. First came a tentative solitary 'peep' as though the creature was finding the right pitch; then a pause; then a more confident 'peep, peep, peep' repeated over and over again like the pulse of the swampland. One-by-one his brethren joined in, the chorus evoking the jingling sound of winter sleigh bells.

"Let's walk over there and try to see one," I said to my daughter.

We put on our rubber boots and took a flashlight. But as we approached the swamp, our boots squishing in the ankle-deep water, the deafening frog chorus abruptly halted, as though the conductor had silenced them with his baton.

"We're scaring them," I said to Ann. "They probably think we're predators coming to catch them."

She thought this over and said, "Dad, how about if we stand still and try to be very, very quiet. Maybe they will start peeping again."

So I turned off the flashlight and we stood still for what seemed a long time. It probably was only a minute or two later that one of little frogs resumed peeping; soon the others joined in. The sound surrounded us; some of it came from almost under our feet. I flicked on the flashlight. The beam lit up a clump of swamp grass; clinging to a blade of grass was a tiny frog smaller than my thumbnail, his throat puffed out in a bubble that seemed as big as his body. The bubble vibrated with each 'peep.'

We bent over and watched, enthralled that such a tiny thing could make such a lovely sound. The flashlight frightened the frogs, shutting down the chorus, so I snapped off the switch and we stood and listened in the dark. The night was made more memorable when, turning to leave, we were startled by a sudden sound overhead—a whooshing of wings so close we flinched and ducked down. A flock of Canada geese, coming in to land for the night, had seen us at the last minute and veered off. Their wings almost brushed our heads!

The geese have traveled on, further north to their summer nesting grounds. But the frogs remain. Although their lusty April peeping is now just a memory, the evidence of their love affairs is here in the countless millions of tadpoles hatching in the marsh pools and slackwater of the creek. One speculates why frogs don't soon overrun the world. The answer became apparent as we watched the activity in the barnyard water trough day-by-day.

Some of the minnows were growing noticeably bigger. The tadpole population was dwindling. Then one day we saw one of the bigger minnows chase a tadpole, seize it and swallow it. It was a story as old as the world. The big fish eat the little fish.

But by now the tadpoles themselves were growing at a rapid pace. Their bodies got plumper and their tails shrunk. Ann and Zachary were fascinated; they kept me informed of the day-by-day metamorphosis.

"Two bumps are appearing at the sides of the tadpoles near the tail. Are those the hind legs?"

And a few days later: "Now they've got their front feet, and the tail is almost gone."

I was as captivated by the process as the children; although I had witnessed it countless springs since boyhood days, the wonder of it was still new. It was almost incredible to watch those little gill-breathing creatures beginning to come to the surface to gulp air as their lungs developed. Here on a Lilliputian scale was evolution recapitulated.

One May afternoon following a warm thundershower, the two scholars hopped off the school bus and made their usual beeline for the trough to view their watery captives. I heard them hoot, "Hey, what happened to the little frogs? They're all gone!"

Sure enough, they had vanished. Then we noticed that the water trough was brim full. The downpour, though of short duration, had raised the water level, enabling the diminutive frogs to climb over the rim and make their escape.

We searched in the tall grass bordering the barnyard and found several of the tiny frogs. They weren't much bigger than the crickets. In the pasture closer to the marsh there were dozens of them. It was a population explosion.

Ann and Zachary whooped their delight and went down on hands and knees to catch the little critters. "I can't believe it!" they exclaimed in unison. "Did you ever see so many frogs in your life!"

Had we not seen the gelatinous egg masses laid in the marsh water in late April, and observed the hatching of pollywogs, this sudden appearance of multitudes of tiny frogs after a thunderstorm would have been cause for wonder and speculation. It would have been easy to believe, as people did in olden days, that the little creatures fell down from the clouds—that it literally 'rained frogs.'

We don't believe such nonsense anymore. We're far more intelligent and sophisticated. We know that it only rains cats and dogs—and sometimes pitchforks and hammer handles.

BEAVERSPRITE

One afternoon in late spring I was driving slowly along route 29 near Dolgeville, looking for a sign tacked to a tree. There it was—a varnished wood sign with the word 'Beaversprite' in prominent black letters. The sign pointed to an unpaved side road that angled north through the dense woods which cover these southern foothills of the Adirondack mountains. Under the name Beaversprite, and printed in smaller letters, was the distance—three miles—and the words 'By Appointment Only', followed by a phone number.

I turned onto the road. Years ago I had heard about an

eccentric elderly woman who lived alone in the woods up here. Her name was Dorothy Richards; she was known among locals as 'the beaver lady.' Once someone who had taken his children to visit her exclaimed to me, "She lets tame beavers come right in her house!"

Solitary living in the backwoods can breed some strange characters. I was curious to meet this phenomenon, and when I did finally visit her back then I was fascinated, both by the beavers and by her story. Almost twenty years had gone by since that first visit; now I was returning because I heard she had written a book about her life with beavers. I wanted to buy an autographed copy of the book from her; and I wanted to see what progress she had made in building a beaver sanctuary.

The narrow gravel road twisted and turned, uphill and down through deep woods. These were second growth maples, oaks and ash. The virgin forest was lumbered off in the 1800's. The cleared land was farmed until the soil wore out in the early 1900's, then the farms were abandoned and the trees sprang up again. It doesn't take long for nature to reclaim her own. The only clues that there had once been farms here were the stone walls, moss-covered, that were still visible in the woods; they had marked the boundaries of long-ago pastures.

As I drove I recalled what I learned about her story on that first visit. She told me she and her husband, Al, bought the place back during the Depression. It was an abandoned farm: fields grown up with weeds, the barn a sag-roofed wreck, the house badly in need of repair, its leaning porch about ready to fall down. Al took one look and said, "I'm naming this place London Bridge."

Nevertheless, its remote location and the beauty of the

woods and the little stream appealed to them. At first they used the place as a summer retreat, then fixed up the house and lived there during the winter as well, commuting to their jobs in Little Falls.

Beavers came to them in an unexpected way. A mammalogist friend from Al's college days dropped in for a visit on a March weekend. He saw that the farm's wild fields, woods and streams were perfect for a project he was doing for the Department of Environmental Conservation. Beavers, he said, by 1900 had become nearly extinct in the state after years of trapping; the state wanted to re-introduce them, so the Department had imported some from Pennsylvania and was looking for suitable areas for their release. He suggested Al write a letter to Albany, requesting a pair of beavers be released on the farm.

That's how the first pair of beavers, whom they named Samson and Delilah, came to them. The beavers dammed up the farm's two small streams—Middlesprite and Littlesprite—creating ponds, building a lodge and over the years raising a succession of offspring. Dorothy, a city girl who had never seen a beaver, became captivated by them and with much patience earned their trust so that they came out of the water to eat out of her hand and even sit in her lap. Eventually she got a permit from the state to keep some beavers in her home so she could study their life habits at close hand.

At first the conservation officials were reluctant. They told her that no one had ever been able to keep a captive beaver alive for the twelve years then thought to be a beaver's life span. But when they came to the farm and saw for themselves Dorothy's unusual rapport with the wild beavers at the pond, they realized her project was not an idle whim and they issued her a license.

Al helped her fix up the cellar of their house as a place for a pair of beavers to live. He built a small pool in the cellar, complete with a wooden shelter for a 'lodge'. The beavers were treated as members of the family, being allowed to climb up the cellar stairs and come into the living room and kitchen during the day.

Eventually they had as many as eleven beavers living in the cellar. The rare privilege of studying beavers at such close quarters and over such an extended period of time enabled Dorothy to observe every aspect of their lives, from mating to giving birth. She kept voluminous notes. She proved that beavers can live more than twenty years, much longer than the ten or twelve year life-span of beavers kept in zoos.

More and more people heard about 'the beaver lady' and came to visit. Dorothy welcomed the opportunity to show how intelligent and gentle the creatures were. She hoped to make people realize that protecting wildlife was a worthwhile thing. My own visit there during those early years convinced me that hers was a noble cause. Now I was paying a return visit to see what progress the intervening years had brought.

The drive through the woods was every bit as lovely as I remembered. In the low lying areas willows and alders, newly leafed in the tender green of May, crowded the road. The perfume of wild apple trees in blossom drifted through the woods, and the white viburnum was in bloom.

The road dipped quite suddenly and at the bottom a wooden plank bridge led over a brook. Just beyond was the brown shingle house in its open glade. I parked the truck in the yard and sat for a moment, struck by the beauty of the scene.

Water gurgled musically in the brook. Ferns raised their

fronds along the bank. I glimpsed a pair of mallards swimming in the pond behind the house.

I got out of the truck, walked up to the house and knocked on the door. After a long minute the door was opened slightly by a rather thin, gray-haired woman who gave me a questioning look. She did not recognize me, of course; I had been just one of many hundreds of past visitors. She looked older, and that was to be expected— I knew she must be in her seventies by now—but age had not dimmed the keen look in her eyes.

"Mrs. Richards, you don't remember me," I said. "I was here to see the beavers quite a few years ago. I've come to see them again and to buy your book. I phoned the day before yesterday."

"Oh? Oh, yes. I'd forgotten." She opened the door wider. "You're a bit early, I'm afraid. The beavers don't become active until closer to evening. But come in. I'll see if I can coax one of them out."

I followed her inside, through the kitchen to the parlor. The room was furnished with a sofa, several chairs, a well-stocked bookcase and a rocking chair. Two windows along one wall looked out on a tiny lawn where chickadees and titmice were busy at bird feeders. The opposite wall was completely lined with windows, separating the living room from what at first glance looked like a totally enclosed side porch.

"That's the indoor pool for the beavers," she explained. "It's twenty feet long and five feet deep. We call it the Y because it looks like the pool in a YMCA. Al, my late husband, and I had it built so we could watch them from our living room. "

"When I was here before, the beavers were in the cellar," I said.

"Oh, they outgrew that. We knew we had to make a better arrangement to show them to visitors. This pool is much larger and the area it's in is more convenient, both for the beavers and for me."

I stepped to the windows for a closer look. The top of the concrete pool was a foot or two lower than the floor of the living room. A door from the living room, and a short flight of steps, allowed access to the area. I could see that the pool had a sloping bottom that rose up out of the water to a dry platform on one end. Scattered on the platform were several short tree branches with the bark chewed off. As I looked down at the pool, all at once the surface of the water trembled; a dark form swam into view, and a beaver emerged, waddled up on the platform and began gnawing on a branch.

"You're in luck," Dorothy said. "The beavers usually sleep until nearly five o'clock in the afternoon. This one is named Eager. She is my favorite."

We sat down. Dorothy proceeded to tell me about her success in establishing a wildlife sanctuary. "Whenever Al and I could afford to, we bought more of the adjacent abandoned farm land. There are now more than five hundred acres in our Beaversprite Sanctuary."

While we were talking, the beaver climbed the steps from the pool area and entered the living room. She rose up on her hind legs and stared at me silently, her front paws clasped in front of her chest. After satisfying herself that I was friendly she dropped down on all four feet and waddled across the wide pine floor boards to the rocking chair where Dorothy was slicing apples in a shallow basket. Eager reached out and accepted an apple and bit into it daintily with her two huge incisor teeth.

Dorothy then showed off a little trick that she had taught the beaver.

"Now, Eager, go and get your pillow and bring it back here."

Eager looked at her intently while she repeated the instructions, then turned slowly and waddled to the opposite corner of the room where there were two pillows on a stool. She rose up on her webbed hind feet, clutched one pillow to her chest with both front paws and walked ever so slowly back. She placed the pillow on the big wicker chair close by Dorothy, then climbed up and sat on it. Dorothy rewarded her with another apple.

We watched the other beavers in the pool for about another hour, a truly memorable visit. Before I took my leave I bought her book, *Beaversprite: My Years Building an Animal Sanctuary*. While she autographed it she told me that Beaversprite Sancruary is one of only three places in the United States or Canada where the beaver is safe. The others are Beaver Lake Refuge in New Brunswick, Canada; and Unexpected Wildlife Refuge in Newfield, New Jersey.

Beaversprite Sanctuary has been placed in a Nature Conservancy Trust so that even after Dorothy Richards' death the beavers and other wildlife will be protected in perpetuity. It goes to show you what one woman and a dream can accomplish.

CALENDAR

The new calendar from the feed company has hung on the wall by the kitchen door since January, a preferential position, because out of all the calendars we get this is the favorite. It's an extra large calendar: a foot-and-a-half wide and three feet high, with date numbers you can read from the far side of the room. That easy visibility gets high marks from me.

But the best feature is that the top half of the calendar has an exceptionally fine picture: a large color photograph of a spring pasture scene — a green meadow with a stream winding through, gray boulders by the stream, a split-rail fence at the meadow edge, and a white blossoming tree leaning over the fence. That photograph has been a balm to many a frozen winter day.

Often at the kitchen door, when getting ready to go out to the barn, putting on boots, parka and mittens, I would pause and gaze at the calendar photograph. It was a little game I had If I stared hard enough I could imagine myself actually inside this scene that bespoke a sunny April day.

I pretended I was out mending fence, and the spring day was so warm I had just stopped to take off my jacket and sit down for a moment on the flat boulder that jutted out of the pasture..

I pretended to pluck one of those new sprigs of green grass and chew it idly while I watched the cloud shadows play follow-the-leader across the vivid green meadow; imagined the shifting patterns of sunlight and shade on the gray wood of the rail fence as the breeze stirred the white blossoms of the wild plum tree. That calendar scene uplifted me; it seemed to say, "What if the snow outside is four feet deep, spring will come, wait and see!"

When I was a youngster, my father had a pasture that looked remarkably like that calendar scene — flat boulders amid green grass, a tiny brook, wild apple and hawthorn shading the fence corner. It was used as a summer pasture for the young heifers.

Since the pasture was at the far end of the farm and somewhat isolated, the heifers did not often see humans and, in such circumstances, could tend to get wild. To prevent that occurrence my father made it a habit to bring salt to the heifers every Sunday afternoon. This kept them tame all summer long. We youngsters loved to accompany him on the long trek to the Calf Pasture; it was like a safari to an enchanted place.

Dad would spread the loose white salt on two or three flat boulders, then call the heifers. They would emerge from the shade of the corner woods and come running to lick up the salt. You could hear the rasping sound of their coarse tongues on the hard rock. Then they would crowd around us to be petted, sniffing our outstretched hands with their wide wet noses, their long, curling tongues rough as sandpaper.

If I stared hard at my calendar picture, I could almost hear the rasping sound of calf tongues licking the boulder. Strange how consciousness can either contract or expand the sense of time. It can think of the past or the future and pretend that they are here and now..

But from this day forward there is no need to play the calendar picture game, no need to pretend, for April is here at last. And here are the white-blossoming trees, and I know these gnarled branches that have put forth new leaves. Here is the tiny brook with crystal water, and I feel its taste. These scents of grass and sunlight, these balmy afternoons when the south wind blows and the heart relaxes — this is the green earth whose power and strength I feel.

It's a warm wind, this south wind, full of birds' cries. It brought the geese weeks ago, returning to their beginnings. It brought the purple finch; a pair is nesting under the barn eaves, hoping to fledge their brood before the return of the swallows who just move right in and take over.

The south wind brought the song sparrow and the chipping sparrow. It brought the tenants of my hayfields — meadowlarks, redwing blackbirds and bobolinks — who pay their rent with a melody. I have seen them all and acknowledged their summer tenancy while I have been at my favorite spring task: mending fence. A dairyman knows, perhaps better than anyone, that good fences make good neighbors.

Yesterday while mending fence in the upper heifer pasture I saw a pair of wild mallards on the pond. When I got too close to them they took off in a splatter of spray, their webbed feet giving them a running start across the water and wings grabbing the air. I wondered if they might be nesting by the pond, so I walked along the shoreline.

I saw some narrow paths among the cattails along the bank, but that might have been muskrats. One of the paths led to a thick brown patch of last year's grass in the bushes — a likely place for a nest.

If the mallards do have designs on nesting there, they had better beware, for the pond has its own 'Loch Ness Monster.' At first glance the thing I saw sticking out of the water seemed to be a dead twig; but when the 'twig' abruptly sank and then re-emerged a few minutes later, I knew it must be a snapping turtle. Woe to the duckling that swims too near those jaws!

Later, while mending fence at the top of the hill, I heard a birdcall that was unfamiliar to me. It sounded like a person trill-ing his tongue — 'prrrt, prrrt.'

The bird was in the maple trees of the fence line, but every time I craned my neck to look, it darted behind a limb. After about half an hour of this hide-and-seek, I saw it — a red-headed woodpecker. He looked like he had a red sock pulled over his head and neck.

If there was any doubt that spring was here, a sound at twilight shattered it. The peepers! Their lusty chorus rose from the marsh, and the stars came out and listened.

THOSE DARN GOATS

"Know what, Mom?" confided eleven-year-old Cheryl, calling in a loud voice as she opened the screen door at the back en trance of the house. "Dad says if he catches those goats up on the roof of the car one more time he's going to get out his rifle." The screen door slammed shut and Cheryl walked through the hall into the kitchen.

"Did you hear me, Mom? Dad said..."

"I heard you, Cheryl," was the patient response. "Now I don't want to hear any more about it."

Prudently, I pretended not to hear. Even I had gotten my

dander up about the goats in the springtime. That was on the first of May when the alfalfa was just starting to green up in the hayfield...hampered by the daily wanderings of the four goats who loved to nibble the new green shoots. That was only typical behavior for goats, of course. Goats hate to be confined; they are wanderers by nature.

In self defense I had loaded up my fencing trailer with supplies — posts and maul, a roll of wire, a carton of electric fence insulators — and driven up to the neighbors.

I didn't want to get angry with my neighbors over anything as trifling as four small goats, and you couldn't ask for better neighbors. But I did have to say something.

"Helen," I said, "you'll have to keep your goats out of my hayfield or I won't have any hay this summer."

She apologized. "Ralph has been meaning to mend the fence in the goat pasture, but just hasn't gotten around to it."

"Look," I said, "I've got all afternoon with nothing to do. I'll be glad to help you tighten up the wires in the pasture and hook up the electric fencer."

Their goat pasture was a quarter-acre grassy area fenced off from the horse pasture. I spent the afternoon helping Ralph mending the fence: tightening wires, hammering insulators on the posts, and repairing the gate next to the barn.

Finally, we hooked up the electric fencer and checked the current. It gave a good tingling jolt. Helen got a pail of grain and coaxed the goats in to the pasture. Then we led each goat up to the fence and touched its nose to the bottom strand of wire. Each one jumped back with a startled 'baa!' at the shock. The four goats circled the inside of the pasture, restless at being thus confined, but afraid to approach the fence.

"There! That ought to fix the buggers," said Ralph.

I climbed on my tractor and drove out of the yard, content at seeing those goats behind a secure fence. Now I could breathe easier, not having to worry about my hay.

Helen's love affair with goats had begun the previous autumn. She likes all animals and once had owned a Guernsey cow which she milked by hand. But Kitten produced much more milk than a family of five could consume. To use up the surplus, Helen became adept at making butter and cheese. But how much cheese can one family eat? She realized a goat would be more practical.

Helen ended up buying three goats — two open does and one pregnant doe. The three goats made themselves right at home during September and October. They loved to eat the falling leaves. For variety they kept the lawn chewed down and pruned the shrubbery — spruce, juniper, dwarf peach tree– pruned them right down to the ground.

They had a fondness for bark. They nearly peeled the young mountain ash tree bare. Ralph saved it just in time by building a chicken wire fence around it.

After observing the diet of those goats, I could understand why goats are pictured in cartoons as eating tin cans. They had to taste anything and everything, from the vinyl top of the family sedan to the plastic wall of the children's swimming pool.

In winter they had to content themselves with mere hay. But with the return of spring Mother Nature rolled up her carpet of snow to reveal the budding grass and leaves. The goats leaped in delight at the smorgasbord.

Their number was increased by one: Amy freshened with a tiny buck kid. The three goats set out to show the new kid around. They nibbled the persistent shrubbery by the house, trotted to the hayfield to check out the budding alfalfa, even meandered over to another neighbor's lawn once a day to taste the forsythia.

One balmy spring day the neighbor left her kitchen door ajar while she went out in the side yard to burn the trash. The goats must have been hiding around the corner of the house just waiting for this opportunity. They pushed open the door and went inside the house.

They nibbled the African violets in the flower pot on the kitchen counter, chewed up a newspaper, and pulled the table cloth off the table. Then they sauntered through the dining room and bathroom, leaving a few calling cards as they went — — the kind looking like brown marbles.

They had a grand time in the bedroom, leaping on the bed. They left a little pool of 'water' on the bedspread.

Fortunately the neighbor has a sense of humor. All she said was "Those goats are the nosiest animals I ever saw. They didn't leave one thing unturned in the house."

Well, the goats wouldn't be visiting anymore, not with the new fence in place.

The next two weeks I was busy plowing and planting corn. I forgot all about the goats until the day I walked in the corn field to see if the corn was up yet.

Goat tracks! I turned and looked across the two intervening fields to the neighbor's goat pasture. The goats were not in the paddock. They were in the far end of my corn field nibbling the tender green tips of newly sprouted corn!

"Out, damn you!" I shouted, and chased them all the way home at a brisk trot.

How the devil had they gotten through the electric fence? I walked the entire paddock fenceline with Helen and soon found the answer. The goats had discovered any number of places in the uneven ground where they could simply wriggle under the bottom wire.

What the devil were we going to do about those goats? If the electric fence wouldn't keep them in, would a woven wire fence placed flush with the ground do the trick? I had to drive to the village that morning, so I stopped our local farm feed and supply store, Case's Mill, and priced some woven wire stock fence. "Do you think a four-foot-high fence would hold goats?" I asked the manager, John Case.

"It might, if you have barbed wire at the top. You know how goats can climb."

"I'm trying to keep my neighbor's goats out of my cornfield," I explained.

"Woven wire can run into a lot of money," he said. "What you might better try is a dose of lead."

I thought I knew what he meant but I asked him anyway.

"What do you mean by a dose of lead?"

"Buckshot."

Well, I wasn't going to try that — not yet.

I drove over and had another conversation with Helen. She came up with a solution that really worked.

"If I tie Amy up in the goat pasture with a long rope, her baby buck will stay near her, and the other two goats won't wander far."

Why hadn't we thought of that before? The goats, definitely herd animals, always wandered together as a group, never alone. And Amy was the instigator of thier escapades.

The goats behaved very well after Amy was tethered; a whole week went by without goat escapes. Success at last! One afternoon we were at Helen's kitchen table having a glass of ice tea, congratulating ourselves on having finally foiled those darn goats, when we heard the screen door of the living room open quietly and heard footsteps on the rug. We paid no attention, thinking it was one of the children.

A few minutes later Cheryl came downstairs, walked into the living room and screamed.

"Dad! The goats are in the house!"

We jumped up and dashed to the living room. Amy was standing on the window seat of the bay window, tasting the draperies; the broken tether rope dangled from her collar. The other three goats were having a grand time vaulting from a stuffed chair to the sofa.

Ralph had the good sense not to get excited.

"Calm down, Cheryl, before you scare the goat and she crashes through the window."

We opened the door and carefully shooed the goats outside and back in to their pasture. We mended Amy's tether.

After we settled back at the kitchen table with our iced tea I said to Ralph, "You really ought to tie some bells on the neck collars of those goats. Then they won't be able to creep up on you like that."

He laughed. "Yeah. A fifty-pound bell! Then I'll take them for a walk along the river!"

He didn't mean it though. And I must admit that those goats are lovable. When they amble up to you so trustingly and gaze at you with their soft doe-eyes, you forget all the mischief they've been in.

SHEEP POWER

As I removed the milker claw from the udder, Gerald reached up to the milk meter to detach the plastic tube with its milk sample, then snapped an empty tube in place. I stepped over to the next cow and slipped the inflations on the teats. Gerry wrote the milk weight on the barn sheet attached to his clip board.

The actions of the milking routine were so repetitious as to be almost automatic. On test day it was agreeable to have someone to converse with, and I listened with interest as Gerry told me about the new lawn mower he recently bought.

"It's got plenty of power," he said. "We need it because we've got nearly an acre of lawn and some of it is on a slope."

I detached another milk unit and waited momentarily while he took the sample.

"Well, then you should be all set for the summer," I said. "Here it is only the third week of May and the grass is about high enough to mow already."

"Oh, we've mowed ours once already. My oldest daughter is twelve and she coaxed me to teach her how to run the new mower as soon as it was delivered."

"There you are! That's one way to get your kids to mow the lawn. Buy a riding mower."

"Right. I made sure she understood all the safety rules though," he replied quickly.

"How many horsepower did you say it was?"

"Ten horsepower."

"I've got a riding lawn mower too," I said. "It does a good job, and it's only one horsepower."

"Huh?"

"And it doesn't burn any gas."

Gerry grinned. "You mean your daughter's horse?"

"Right!"

"I saw him eating grass on your front lawn as I drove in this morning. Aren't you afraid he might wander off and get on the highway?"

"Oh, no. He's very reliable. We let him graze on the lawn for a couple of hours early in the morning and then put him back in his paddock for the middle of the day. We let him out on the lawn again for a little while at suppertime."

"Don't you have to mow your lawn at all?"

"Last year I only mowed it three times, and that was just to trim the tall bits of grass here and there that the horse missed. But this year I won't have to mow it at all."

"Why not?"

"Because the one-horsepower riding mower is being replaced by" — here I paused for dramatic effect — "four sheep-power."

Gerry grinned again. "Do you let those sheep out on the lawn too?" He was referring to the pet sheep in the paddock next to my machinery shed.

"Sure. They do an even better job than the horse. They eat the taller grass and weeds that the horse is too finicky to eat. Not only that, but they eat the grass right down close so that it looks like a regular lawn mower has done the job. They even trim the grass under the peonies and shrubbery."

"I don't suppose they leave big calling cards like your horse does? You know what I mean, those clumps of horse manure."

"Not at all. The sheep berries are small enough to get lost in the grass. Look at the benefits I've got: my lawn mowed for free, and fertilized at the same time. That's how the barons and earls in England keep their estates trimmed. Could I sell you a sheep?"

"No thank you. One dog and one cat is enough livestock for me." Gerry took another milk sample. "Where did you get those sheep anyway?"

"They were a project for my daughter. She got two lambs last year. They grew up and had lambs of their own this year."

"So you doubled your flock in one year. If that keeps up you'll be in the sheep business instead of the dairy business."

"I don't think I would mind that a bit. There would be considerably less work. And I bet there would be more money in it if a ewe could fatten two lambs every year. Have you ever bought lamb chops or leg of lamb at the market?"

"Are you kidding? We can't afford those prices."

"That's what I mean."

"Well, what are you waiting for? The dairy business is heading downhill. Get more sheep. You can even sell the wool."

"Speaking of wool, you should have been here last week when we sheared our two ewes. It was the funniest thing; you would have died laughing."

I told him how naked the sheep looked after they were shorn of their thick fleece; they looked almost like skeletons.

"And you could tell they felt naked too," I said. "When we put the two sheep back in the paddock, they didn't recognize each other; they kept backing away in alarm. Then they charged each other head on.

"And you should have seen the lambs! They didn't recognize the two skinny ewes who were their mothers. They took one look, uttered a startled 'baa' and scooted away in fright. It took them an hour, with a lot of sniffing and smelling before they realized these really were their mothers."

"I wonder what it's like when a commercial flock of several hundred is sheared. Do you suppose it's the same response?"

"If it is, it must be a circus," I said.

"Gosh, but those two baby lambs you've got out there in the paddock are cute," Gerry said. "Would you mind if I brought my daughters here to look at them? They would fall in love with them."

"And then you would have to buy a sheep of your own!"

"On second thought, I don't think I'll let my daughters see those lambs!"

HARDWOODS in MAY

Springtime at last! What a wonderful feeling to go outside in early morning and see new hues appearing on our hills and in the woods where lately there was only a monotonous winter white.

The hardwood trees at this earliest season treat us to a double display of color—first gold, then green. Just now in the fullness of May as they put on their new leaves their hue is a gossamer green. But earlier, at the end of April before their true leaves unfolded, was their flowering time of shimmering gold.

The blossoms of the sugar maples were tiny yellow bells,

borne in clusters on the twigs of every bough. The majestic trees thus arrayed seemed covered with a golden veil; from a distance they appeared to be in tiny leaf. But a close inspection revealed the leaf buds still tightly closed, although swollen and ready to burst open. The blossoms of the ash and elm were similar to the maples.

Nature's schedule of priorities for hardwoods seems to be reproduction first, leafing-out second.

This morning as I drove the tractor down to the village to have some repairs done, I felt the exhilaration that spring time always brings. It was a leisurely ride on our country road which for about a mile winds through a dense forest of pines and hardwoods. The beauty of the new leaves was glorious.

The oaks, hickories and maples were a tender shade of green, contrasted against darker pines. Although the morning sky was overcast, the delicate foliage lent a luminescence to the landscape, as if the sun were shining through mists.

The creek, which the road follows descending to the village, was rippling over boulders and foaming down small waterfalls, its flow only slightly diminished from the swift current of early May. Along one stretch of bank the wild day lilies, whose gnarled tuberous roots have survived the scouring action by floating ice during countless winter thaws, had sent their sword-shaped leaves up waist-high already. Their growth rate is prodigious; I'll venture it's at least an inch a day. By the first week of June they will be opening their trumpet shape orange blossoms.

This portion of the road within the woods is such a pristine area that I always drive through slowly; today, being in no hurry, I parked the tractor for a few minutes where the creek comes close to the road.

The narrow space between pavement and the water's edge was covered by a delicate green carpet of maidenhair ferns. From the woods, tall maple trees leaned far out over the brook. At the tip of one high overhanging branch an oriole had suspended its basket nest; from nearby its warbling notes cascaded a melody as sweet as an oriole can sing.

Under the woodland trees I could see that the forest floor, which in April had yet been a monochrome of dead brown leaves, now was green with flowering plants. Solomon's Seal grew in dense clusters, its arching leaves like miniature palm fronds. Wood sorrel, columbine, wild ginger, all were in bloom.

How nature rushes to get things accomplished before the longest day of the year arrives in June! Deep in the snow of January I forget how delightful May can be. I am refreshed by the sight of her inexhaustible vigor.

That night after milking and after supper, I took the tractor to the top of the creek pasture hill to mend the fence along the woods. This was the last portion of broken fence remaining to be mended—fence that had been damaged by falling limbs during winter storms—and I had about an hour of twilight by which to work.

Although the morning and early evening had been cloudy and overcast, the weather cleared before evening and the sun went down in a blaze of color. The robins gave their evening serenade in that half-hour between sunset and the first deepening of twilight, while I was busy replacing fence posts. The night was warm, and as I stopped to wipe the sweat from my forehead I heard the faint but unmistakable chirping of a cricket, the first one of the season.

As I set the new fence posts and drove them home with

swinging blows of the maul, the sound carried through the ranks of trees and echoed back. Now twilight was deepening; the robin serenade stopped abruptly, as though someone had flicked a switch. Darkness swiftly increased, and I hastened to tighten the wires while I still had light enough work.

Now I had a new accompanist. Nearby in the pillared dark of the woods, a hermit thrush began to sing. His flute-like notes, two long and three short, were as pure and sweet as they could be. It almost seemed like a sacrilege for me to be intruding on his melody with the metallic sound of hammer on staples.

Anyway, it was getting too dark to see what I was doing. I was swinging the claw hammer more by guess than by sight. Only one blow out of five hit the staple. Time to go home.

Before I turned the ignition key to start the tractor, I listened a few more minutes to the song of the thrush. At that moment I wished it could be forever springtime, with me lingering on that hillside at the edge of the woods as the stars, one by one, began to twinkle in the evening sky.

HORSE TRAINING

Who trained whom? Reflecting on that unspoken question, I watched the three-year-old chestnut filly stepping smartly across the farmyard, my fourteen year-old daughter in the saddle. She waved as she passed by, calling out that she would return in time for her chores at four o'clock.

The horse clearly was looking forward to her daily excursion. Her neck was arched, her ears pricked forward, no hesitation in her movement; every muscle rippling beneath the sleek brown hide seemed to say go! Barely a month earlier her training had begun in earnest. As is the case every spring my own farm work at that time was accelerating by leaps and bounds; and we had two young horses that I never seemed to find time to train; a frustrating situation if there ever was one.

My daughter's first thought upon getting off the school bus every afternoon was to ride her gelding, Andy, for an hour before she had to report to the barn for her chores at milking time. I couldn't begrudge her that free time after being cooped up in classrooms all day; yet the two young horses needed to be trained, and one or both of them sold.

Last summer, over a period of only three or four days, we had broken the two-year-old filly, Diamond, to the bridle and bit, and even drove her with the long lines. But then work intervened and her training was suspended.

"You've got to resume the training all by yourself," I said to my daughter. "I just do not have the time between corn planting and haying, and a million other things. We've got to get Diamond and Prince trained and sold. Seven horses are just too many around here!"

She looked crestfallen. "But then I'll never have time to ride Andy," she protested.

"Yes you will. But you've really got to begin working with Diamond a little bit every day, even if it's only fifteen minutes at a time. A brief lesson each day is actually better for a young horse than a long, boring lesson once a week."

After only a bit more grumbling, she agreed.

The first day's lesson was spent merely grooming the filly, getting her used to being touched and handled again. Next came more sessions with the long lines. I could tell how the training was progressing just by listening to the conversation between girl and horse: "Whoa! Giddyup! Back up! No, not sideways.

Back. That's the way! Steady now. Good girl! Goood girl!"

Then came sessions of putting the saddle on and having the horse get used to the cinch being tightened around her belly. Finally the day came when my daughter called, "Dad, will you come and hold Diamond while I mount up?"

A horse may go through all the preliminary lessons with flying colors, but you never know how it will react the first time you actually get in the saddle. It's always a tense moment.

Diamond barely flinched as this new weight was put on her back. Would she buck? But no, she merely turned her head to look as Ann seated herself in the saddle; and she moved off smoothly when Ann gathered up the reins. Once around the yard was enough for that momentous occasion.

The next day she was ridden a quarter of a mile up the side road. By week's end she was doing so well that I allowed Ann to ride her a mile to her girlfriend's house at the far end of our road.

A few training lessons later Ann decided to switch from the English saddle to the heavier western saddle. Then, of course, the filly had to be taught to neck rein. This took the better part of a week.

It was gratifying to see how well the filly took to her training. She either was waiting at the pasture gate each afternoon or, if grazing, came readily when Ann called to her. The filly enjoyed being tied with the halter while she was being groomed and stood quietly while the saddle was put on and tightened. She looked forward with eagerness to the jaunts over the hill to the girlfriend's house. The girlfriend owned an Appaloosa mare, and the two girls rode the trails together.

It probably was just as well that we waited another year to

train the filly. She outgrew the skittishness and scatter-brained quirks of the two-year-old and was ready to settle down.

Seeing my daughter astride the filly, the two moving smoothly as one whether at a walk, a trot or a canter, I could not help reflecting on how training affects both the horse and the trainer.

The horse must learn to respond to the bit; the trainer must learn that the bit is a guide, not a weapon. In time, the horse will respond to the gentlest pressure. Such a horse will be said to have a 'wonderful mouth' and the rider will be said to have 'wonderful hands.'

Everything that training consists of is foreign to the nature of the horse; and the horse, being such a powerful animal, could effectively resist being trained at all. Therefore the trainer must learn to use intelligence, not muscle, in the training. Patience is everything. I noted its progress in the rapport that developed between horse and girl.

The filly is well on the way to becoming a fully trained horse. The girl is well on the way to becoming a self-reliant woman. Today, returning from her ride, Ann declared, "I'm proud of this horse, if I do say so myself!"

Diamond tossed her head as if to say, "I'm proud of this girl!"

GUS

Gus stood in the open doorway of the barn and watched the hired boy dash down the road toward the village. Sammy had a stride like an Olympic runner; his legs could eat up the ground faster than any fifteen-year-old-boy he knew. Of course, Sam had an incentive: the school baseball game was scheduled for 7 p.m. and he was the starting pitcher.

Sam had rushed through chores, grabbed his baseball glove and sprinted off. Gus offered to drive him, but Sam refused, saying "Naw, it's only two miles; it's all downhill and it keeps my legs in shape for running bases."

Gus leaned on the barn broom, watching the figure disappear around the bend in the road. Never did see a kid so nuts about baseball. No, that wasn't true, Gus mused, remembering he had been that way himself as a teenager.

He recalled those warm spring afternoons when he hurried home from school and rushed through chores — he and his brother, John, each had to milk four cows by hand in those days — before their father would allow them to race to the ball game in the village.

Where had the time gone? John was dead these two years now.

Gus had nourished a secret hope that one of his sons, or one of John's, would be interested in the farm, but none of them had. So Gus carried on alone.

But the years were beginning to tell, and he wondered how much longer he could carry the load. He knew he had to think seriously about retiring. In few more months he would be seventy. Think of it! It didn't seem possible that nearly fifty years had passed since he took over the farm. Fifty years! What a long time to be milking cows!

The thing was, Gus knew he had to retire, but he wanted to live out his days on this old farm — slow down a bit, yes, quit the daily grind of milking — just raise a few heifers and bale some hay to sell.

But would that generate enough income to pay the damn land taxes? On the other hand, if he sold the farm, the damned government would take half in capital gains taxes. He had always thought of his equity in the farm as his retirement nest egg. It didn't seem to work out that way after all.

He turned it over in his mind from time to time: there's the rub — you work all your life but you can't even retire on your own acres. You have to get out of the way so the land can be taxed.

Gus shifted the broom in his hands and commenced sweeping hay to the cows. His were the big, rough hands of a dairy farmer. Think of the work they had done! He wished he had a nickel for every bale of hay they had lifted, or every calf they had delivered. There was a strong maternal side to farming; you were always bringing new life into the world—a new calf, or a new seeding.

It seemed his whole life had been a partnership with cows. He had taken good care of them, and they had responded by being good to him — those amazing mortgage lifters! The old fairy tale of Rumplestiltskin spinning straw into gold made him smile. Cows are far better at alchemy. Cows spin hay into golden milk. Gus put the broom away and walked outside.

It was one of those beautiful evenings in early May, the kind of spring evenings when he suddenly knew why he put up with so many cold, bleak, windy winter days—they were all necessary for the creation of an evening such as this.

He let his senses drink in the scene. Along the brook, the willow was bursting into leaf. There was the smell of new, green grass in the air; and the soft evening air itself felt warm on his face. The barn swallows were twittering as they glided in and out of the hayloft.

Gus felt his blood surge within him. Yes, and what does the Good Book say? Till your land for six years, and on the seventh let it rest and recover. The Sabbath. Sabbatical. Was it only for teachers and professors? Did not the scriptures speak to those who till the land?

And after seven times seven it will be your fiftieth year.
Then take your trumpet and blow, for this is your jubilee.

COW in the WELL

It was one of those days on a dairy farm that pop up every once in a while, the kind that made me wish I had stayed in bed.

It was a blistering hot afternoon, unusually hot for May, with the thermometer climbing well above ninety degrees. I was planting the last of the field corn on the high level crop ground behind the heifer barn.

Back and forth across the plowed field I steered the tractor, row after long row. The planter trailed behind, stirring up a small cloud of dust that hovered in the air. From my seat on the tractor I could look across our little valley to the wide sloping meadow we call Rivenburg Hill; the dairy herd had been put there for the day to graze the lush ladino pasture. But there was no sign of them on the broad green hillside.

Most likely they had grazed for two hours in the cool part of the morning and now must be lying down in the shade of the woodlot bordering the pasture. I envied them the shade; I didn't relish being out in the scorching sun either, but I had no choice. The corn had to be planted.

At least the cows had water that came from a shallow well in the front corner of the Rivenburg pasture. Over a hundred years ago a spring had been dug out, enlarged, and lined with field-stone, making a well four feet deep. From the well a half inch galvanized pipe ran downhill fifty feet and fed a water trough.

The cows had water, but I didn't. My thermos of cold water had run out long ago, but I was darned if I would interrupt planting to make a long trip back to the farm house for water. I couldn't spare even the ten minutes of time it would take to do that. Thunderstorms were forecast to roll in before evening, so I had to finish planting this corn by milking time.

Hot sun, the acrid smell of fertilizer, the bitter odor of fungicide from the seed corn: all this was part and parcel of planting time. The dust continued to rise from the ground behind the planter's press wheels as the drive chain went clickety-click. It didn't seem possible that my parched mouth could get any drier.

At quarter to five Alan, the hired boy, went to fetch the cows from the pasture and bring them to the barn for the evening milking. I finished planting the last rows of corn and headed back to the farm yard with the tractor and planter.

The cows already were coming up the lane to the barnyard. Now I would have a few minutes to wash the dust off myself and quench my thirst. There might even be five minutes to relax before I started the evening milking..

But it was not to be. No sooner had I settled down with a glass of iced tea than Alan rushed in the house.

"There's a cow stuck in the well!" he said.

"Oh, no!" My heart dropped as a dire thought leaped in my mind. "Is she dead?"

"No. But she's really wedged in the well, stuck with all four feet and she can't get out."

Regretfully, I gulped the iced tea and did some quick mental calculating. "Let's put the three-point hoist on the rear of the tractor. Get the chain and the coil of heavy hay rope. We'll have to pull her out."

It took less than five minutes to assemble the gear and drive the tractor to Rivenburg Hill. And there was the cow — it was Juniper — huddled in the narrow well like a cork in a bottle. Well, almost, though not quite as tight.

Her head was out of the water, resting on the stone rim of the well. She mooed disconsolately. Alan said if it hadn't been for the mooing he never would have noticed the cow, hidden as she was by the bushes that screened the well.

"Juniper, how in the world did you do it?" I said.

She mooed again; she was quivering all over in distress. I leaned over the well beside her and began sliding one end of the rope over her back. Then, carefully, I lowered myself in the water beside her and groped under her belly for the rope. I didn't want her to start thrashing around and pin me against the stone with her legs. I talked soothing words while my hand searched for the end of the rope dangling somewhere in the water on the other side of her chest in the narrow space between the stones and her rib cage. "Juniper, it's a long way under your belly," I gasped. My arm wasn't long enough to reach the rope. I pushed more slack over her shoulder and groped again under her belly.

"Man, this water's cold! Now I know why you're shivering."

At last I felt the end of the rope; I pulled it and cinched it around her chest. We lowered the hoist and fastened the rope to it. Then I climbed on the tractor and eased the hydraulic lever.

As the hoist lifted, Juniper rose slowly out of the water. Her front feet cleared the wall, felt the ground, and she scrambled out on her knees. Shakily, she stood up on all four legs.

"Easy now, girl," I said as we untied the rope from her chest. She dripped cold water and trembled violently as her muscles tried to generate heat.

Other than chafe marks on her hocks, and a bruised teat, there were no serious injuries. She shivered all the way home. Her barnyard mates eyed her with mild interest as she shouldered her way among them.

I looked at the herd and wondered which of the cows had shoved Juniper into the well, because that's what must have happened. The day having been so unusually hot, they drank the water trough dry and then went to the well searching for more water. With a group of cows pushing and crowding, one of them was bound to get shoved in the well. Cows have no manners.

However, it wouldn't happen again. We put a new cover of heavy planks on the well. Our minds will be at ease. All's well that ends well.

SOMETHING of INTEREST

When I bought the farm and gladdened the hearts of the bank directors with a huge mortgage, a well-meaning person warned me: "The interest will kill you."

Sometimes it seemed that way when I pondered that of the monthly mortgage payments only one-quarter went toward reducing the principal, while three-quarters was interest. This discouraging fact of financial life can tarnish the brighter side of farming, if the farmer tends to be a worrywart.

One soon learns there is no percentage in worrying. Let the moneylenders have their pound of flesh. Who cares if the mortgage interest piles up in that darn bank book? In our house we started a little book of interest that compensates for all that.

We made our book of interest from a child's school notebook. Its use as a book of interest came about quite by accident.

One dismal spring day when just about everything that could go wrong on the farm did and I came home for supper in a black mood, I saw something in the back yard that made me forget my troubles. Wanting to share it as a sort of game, I spied my daughter's school notebook on the table, and on impulse opened it up to a blank page and wrote: 'Look at the top pipe of your swing set on the lawn. You will find something of interest there.'

When my eight-year-old daughter found the note she at once was intrigued and dashed to the back lawn. After a minute of scrutiny Ann found the 'interest'— a pair of house wrens were making a nest in the top pipe of her swing set.

Naturally, she couldn't wait to find something of interest herself and write her own entry in the book.

A few days later I read this brief sentence in her handwriting: 'There is something of interest in the barnyard water trough.'

When I brought the cows in for milking, I walked over to the trough and looked in. A swarm of black pollywogs dimpled the surface as they turned tail and dove for deeper water. As Thoreau said, "Some circumstantial evidence is very strong, as when you find a trout in the milk." In this case it was pollywogs in the water trough, and the evidence strongly pointed to the perpetrators. Ann and her friend, nine-year-old Zachary who lived just up the road, had been seen catching pollywogs and minnows in the creek, using an old coffee can. Very interesting!

The book of interest soon became a game, with each family member vying for the most interesting entry. We became aware of more things that ordinarily we would have taken for granted.

The next entry I made read: 'If you look over the railing of the bridge you will see something of interest in the creek.'

The other members of the family discovered what it was in short order—a brown trout that lived in a shallow pool under the bridge. We all soon began tossing items of interest to him—grasshoppers and earthworms. He became our pet trout. We looked for him every time we took the cows out to pasture.

Another entry in our interest book said: 'Stand by the barnyard fence when it starts to get dark and you will see something of interest at the feed bunk.'

That night at dusk I was watching. The barnyard was deserted and quiet. It got almost too dark to see the feed bunk when suddenly I discerned a furry form with a masked face climbing over the far edge of the feed bunk, scurrying over to the cross conveyor and clambering through that tunnel: destination silo.

The masked bandit was followed in quick succession by three smaller furry critters. It was a family of raccoons headed for an evening appetizer of corn silage on the half shell.

One raccoon apparently thought the covered cross conveyor was a hollow tree. He must have been snoozing there at 6 a.m. when I flicked the switch that sent the chain and paddles surging forward. When his furry form scuttled out at my feet, we spent a few interesting seconds trying to get out of each other's way.

An April entry in our interest book read: 'Go outside after

supper and stand still and listen. You will hear something interesting in the alder bushes of the Brushlot Pasture.'

After supper we went out on our back lawn where the grass was brown and soggy from the melted snow. Twilight deepened, and soon enough, we heard a sound coming from the alder thicket—"peent, peent"—repeated methodically about every five seconds. We named him 'the bird that goes beep in the night.' Sportsmen know him as the woodcock or timberdoodle.

By any name, his mating call is sweet music to our ears. He is a true harbinger of spring. After a few minutes of his 'peenting', which he does while perched on the ground, this bird will fly straight up in the sky, so high that he is a mere speck. Then he will descend in a slow spiral, all the while making a sweet twittering sound to impress his mate, hidden in the alders. After landing on the ground, he begins 'peenting' again for another five minutes or so; then he flies high up in the sky again and repeats his twittering, like a skylark. Over and over, and yet over again he repeats this performance, and we have a grandstand seat. This aerial display is impressive; on moonlit nights he will keep it up until nearly dawn — and it's all for love!

A winter entry in our interest book once suggested: 'You will feel something of interest in the hay in front of Icicle's stall.' What I felt was four newborn kittens delivered by the mother cat, Tina. The wonder of it was that Tina elected to have her kittens right next to the entry door of the stable — she who always took great pains to hide her kittens in the hayloft. She must have finally determined there was an advantage in being closer to something she had a decided interest in — the milk in the cat dish by the door.

JEFFERSON DODGE

Busy though I was with mid-morning barn chores, I noticed the late model car drive in my farmyard and surmised at a glance that it was a salesman. For a fleeting instant I had the urge to imitate Byron Klump, a local farmer. Byron's defense against salesmen was to hide in his silo room until they gave up looking for him and drove away.

But I went on with my chores and at the same time got prepared to listen to the sales pitch of the earnest young fellow with the green cap. He introduced himself as the territory representative of a brand new feed company in the area. I gave him my standard response to new feed salesmen.

"I'm quite satisfied with the grain that I'm feeding now, but leave me your price list and I'll keep you in mind."

He jotted down some figures on the back of his business card and handed it to me as he left saying, tongue-in-cheek, "Don't throw it in the trash barrel until I drive out of the yard."

I smiled and looked at the card. It had the feed company logo and his name, Jefferson Dodge. I tucked the card behind some others by the door jamb and forgot all about it. I didn't think I'd see him again. After all, three major feed companies already serviced this area.

But two weeks later he stopped in again. He had the current price list. His company's dairy pellets were edging downward two dollars a ton, but still not enough to tempt me. He was personable and easy to talk to, and knowledgeable about purebred Holsteins, so we talked cow talk for a while. My curiosity got the better of me and I asked what success he was having getting new accounts in the area.

"I've only gotten one account so far," he said. "It takes time to work into a new territory. The first few weeks a salesman has to sell himself. After the farmers get to know him, then he can begin to sell feed."

It was about a month later, with the price spread now five dollars a ton, that I ordered my first load of feed from him. A few days later he stopped in to see how the cows were doing.

"Do they like the pellets?" Jeff asked. "Are they holding up on their milk?"

I assured him the cows were doing okay, except for a couple of finicky eaters that were slow to adapt to the slight change in taste of the new pellets.

"And how are you doing?" I asked. "How many hundreds of tons of grain have you sold this month?" I was beginning to take an interest in the uphill battle of this neophyte against the entrenched grain giants. It was like David pitted against Goliath.

"I had a real good week," Jeff answered. "Picked up one new account, and two others seem definitely interested. I feel really good about it." The enthusiasm showed in his voice.

"Nothing succeeds like success," I said.

"That must be true. This one new account, and two others on the brink have given my self-confidence a boost."

On his next regular visit, the enthusiasm was still there, although Jefferson had gained only one of those two farmers that had been 'on the brink.'

"I don't know what to do about that other guy," he said. "I took forage samples for him last month and ran a computer program to balance his ration. He says his cows have never milked so good as they have since he began using my program, but he is still feeding the competitor's grain!"

"Keep trying!" I said.

On his next visit, Jeff told me about another account he got.

"Boy, I fell right into it!" he said. "I stopped at this man's farm just after a competitor had delivered a load of grain. The farmer's grain bin was in one corner of the haymow that was on the banked side of the barn. All the truck driver had to do was back up to the haymow door and stick one flexible tube into the porthole of the bin. No long pipes to hook up or anything."

"So?"

"Well, the truck driver started blowing in the grain. Then

he went and sat in the cab while the truck finished blowing in the feed. He didn't realize that, towards the end, the flexible tube whipped free of the porthole and blew pellets all over the haymow floor!"

"What a mess that must have been!"

"It was! And the worst part of it was that the truck driver never even told the farmer about it. He just got back in his truck and drove away."

"Whew! I bet that farmer was happy!"

"He was spitting mad. He called up the plant, and they told him they would send him five hundred pounds of pellets extra on his next delivery at no charge, to make up for the spillage."

"But that didn't soothe the farmer?"

"No! The whole thing was the inconvenience of cleaning up the mess, of the pellets blown all over the bales of hay." Jeff chuckled. "I told the farmer that if one of our trucks ever did that, I personally would come and clean up the mess myself."

"And so, as a result, you've got a new account?"

"Yes, I do!"

On his next regular visit, Jeff had yet another success story to relate.

"A farmer phoned me at nine o'clock the other night and ordered ten tons of dairy pellets." Jeff went on to explain that two weeks previous he had taken forage samples for the farmer and sent them to the lab for analysis. When the results came back just three days ago, Jeff dropped them off to the farmer. The sales rep for another feed company also had taken some forage samples for the farmer at the same time but was slow in getting the results delivered. "The farmer told me over the phone that he was sick and tired of waiting, so he was switching to me!"

"The early bird always gets the account!" I said.

"Then the farmer apologized for calling so late at night. I told him he could call at midnight to order grain; it wouldn't bother me a bit!"

As the weeks went by, Jeff slowly picked up more accounts. But that early farmer who had been 'on the brink' still was uncommitted, and that continued to bother Jeff.

"I don't know what to do," he confessed. "Here he is with his cows milking so well on the program that I wrote for him three months ago, and yet he is still using the competitor's grain."

"Why do you suppose he won't change?" I asked.

"Well, for one thing, he and his wife are very religious. I asked him this last time if he thought my company had good grain, and he admitted that it was. Then I asked him if he was entirely happy about the computerized feeding program I wrote for him, and he said he was very happy with it. Then I said that maybe I was pestering him too much, and would he rather I didn't stop in so frequently. He answered that, on the contrary, he looked forward to my visits. Then I said, by golly, what do I have to do to get you to try my feed?"

"What did he say?"

"He answered that his wife just didn't feel at 'peace' about making a change just yet."

"So?"

"So then I asked him, 'What do I have to do to make you both feel at peace? Would it help if I prayed?'"

"What did he say to that?"

Jeff smiled and answered, "The farmer brightened up at

that and said, 'You know, it might help if you did pray.'"

"Well, are you going to pray?" I asked.

"You bet I am! I'm going to pray tonight like I never prayed before!"

OH, BEANS!

The square white envelope was just thick enough to be intriguing. It was from my sister, Rose, and it was a birthday card, of that I was quite sure, even though it was a week late.

I tried to guess what else was in the envelope. It was too thick and bumpy to the touch to be photographs — perhaps it was a handkerchief. She often enclosed special little things in her letters; once it had been two tea bags of an unusual herb blend.

I gave up guessing and opened the envelope; out came a

package of garden seeds — green bush beans. It was a vegetable seed greeting card! A real package of beans seeds with a picture of a bean plant on the front with long green bean pods. At the top of the packet was printed the greeting: "Oh, beans! I forgot your birthday!"

The card was printed by a novelty greeting card company in Oregon. Inside was an appropriate verse and a reminder that the seeds could be planted in the garden and yield a remembrance of the event months later.

I learned that this greeting card company uses packets of a wide variety of garden seeds on its novelty cards. But this one had special significance. In the note below her signature my sister explained why: "Dick, the moment I saw this card I thought of Dad, and I just had to send it to you!"

I smiled as the memories came flooding back, bringing a warm feeling to my heart.

"Oh, beans!" was one of the two cuss words our father used. The other was "Oh, sugar!" They were Dad's five-letter words, opposed to the four-letter cuss words most people use.

Dad took great pains to explain to us youngsters that swearing was a bad habit and an abomination. If something went wrong, provoking one of us pre-teenagers to spout, "Oh, hell!" or "Oh, —!" Dad would patiently take us aside and explain, "It's just as easy to say 'Oh, beans!' as it is to say something bad. You will feel better by starting a good habit and people will respect you for it."

We tried, but secretly we thought it was silly to say a thing as weak as 'Oh, beans!' when 'Damn it!' sounded so much more forceful and grown-up. Children are always anxious to try out the new words they learn from their schoolmates.

Dad wasn't the kind of person to say one thing and do another. To this day I can only recall one instance of hearing him swear — not that there weren't provocations. In retrospect, I am convinced his self-control was amazing. When the cows got in the corn field, it was 'Oh, sugar!' When the truck fender got smashed it was 'Oh, beans!' Those same kind of frustrating happenings tempt far stronger language from me today; but then I try to remember Dad's admonition.

Oh, I try, and now and then my self-control doesn't waver and I manage an 'Oh, beans!' in spite of the anguish.

But I always thought 'Oh, beans!' was an expression known only to our family. So it was with a little thrill of pleasure that I read the greeting card from Oregon with its 'Oh, beans! I forgot your birthday!'

Self-control is something you don't think about too much until you become a parent yourself. Only then is it borne upon you how important it is to set a good example for your children. Only then do you truly appreciate the efforts put forth by your own parents in all the years they were trying to set a good example for you.

Self-control is called upon quite often when working around dairy cows. These docile creatures can be irksome at times. When dealt the blow of an unexpected kick, a farmer is hard put not to swear and kick back. It helps to have a Rules of the Barn card framed and hanging on the barn wall. I have such a card; on it is the famous quotation by W.D. Hoard that begins: "The rule to be observed in this stable at all times is that of patience and kindness toward the animals...". Quite frankly, I could use half a dozen of those cards tacked to the wall at various spots throughout the barn to serve as gentle reminders to myself.

That I am not alone in needing such reminders was borne out in an anecdote related to me by an Amish farmer friend. Amos admired the Rules of the Barn card so much that I gave him one of the cards to nail up in his own stable.

"But even so, I lost my temper and hit a cow," he told me. "My twelve-year-old son spoke up quickly, saying 'Remember, Dad — 'Rules of the Barn!'"

After I plant that package of seeds I think I will tack up the empty packet alongside the Rules of the Barn.

It will serve as a gentle reminder that swearing is bound to offend the cows.

I wonder if Dad realized the significance of his expression 'Oh, beans!', and what a lasting impression it would leave. I still can see that weathered face with the hooked nose and the twinkling eyes looking earnestly into ours as he patiently instilled his wisdom.

It was the ritual of taking the time to explain, always with utmost patience for the small human person involved, that imbued those two words with such meaning. Gentle language can still be strong. If nothing else, it taught me never to underestimate the power of a good example.

MAPLE BEES

It's May, and I spent a soul-satisfying morning on the Creek Pasture hill laying out temporary fence for our intensive grazing while spring all around was bursting at the seams. In full bloom was the shadbush, the earliest flowering shrub of the countryside; its ivory blossoms conspicuous against the reddish-green of the opening leaves made islands of white in the hedgerow and along the woodlot border.

If only I can remember the location of these shrubs come late June when the fruit ripens like miniature cherries, I might be able to pick enough for a pie. None better!

As I unrolled the polywire my thoughts turned briefly to the previous evening when Alice and John, neighbors newly moved from Long Island, had visited. I smiled as I remembered what they told about city friends who had stayed with them for the weekend and exclaimed with amazement, "How can you live here in the sticks with no cable TV?" I smiled again as I recalled a deft definition of TV — 'chewing gum for the eye.'

Who needs TV when spring spreads a feast for the eye? On a day like this I had no thought but to erase all thoughts and let nature bombard my senses. On such a day I could understand why the Captain of the Pequod said "Oh, but Ahab never thinks; he only feels, feels, feels; that's tingling enough for mortal man! to think's audacity. God only has that right and privilege."

Under a maple tree in the hedgerow I was fastening an insulator to a fencepost when I heard a deep humming overhead. Looking up, I expected to see a swarm of honeybees clustered on a branch. Since the leaf buds had barely begun to unfold, the tree was open and I could see all the branches clear to the top, but no swarm was visible. I walked a few steps to the next tree which was an ash; perhaps the swarm was there. But no, and now the humming was fainter.

I walked back to the maple and the humming grew loud. The whole tree seemed to vibrate with the sound. Puzzled, I looked up at the branches intently. The maple was in blossom, its inconspicuous yellow-green flowers covering the whole tree like a halo of gossamer. And now I could see on every tiny flower, on every tiny twig a wild honeybee was working. The entire work force of the hive must have been in that tree.

I wondered if they were gathering nectar or pollen. Probably pollen, their source of protein for the brood back in the hive. The bees would be ravenous for it after the long winter. It's a barren time for bees, flower-wise, from snowmelt in March to the first dandelions in mid-May; although they did have some pollen from the pussy willows in April.

Recently I read about work being done with honeybees at the Carl Hayden Bee Research Center in Tucson, Arizona. They are putting bar codes on bees — stripes like those found on packaged foods in the supermarket. They are the world's smallest bar codes.

Nine stripes, less than one-tenth of an inch long, are glued to each bee's back. These bar codes weigh less than 20 millionths of an ounce, or about one-twentieth as much as the nectar and pollen normally carried by bees on their foraging trips from the hive. An electronic bar code reader at the doorway of the hive records each bee's exit and entrance and then stores the information on a computer.

Imagine, a dossier on each bee! A bee FBI! They intend to sort the lazy bees from the busy bees, and pick parents for future generations of super-busy bees.

Man is never content with nature. He feels compelled to meddle. Well, these wild honeybees on this maple tree are free from tinkering.

The morning slipped by, made memorable by the sighting of the first indigo bunting returned from the south —a bird to out-blue all bluebirds — and hearing the rapturous warble of the first oriole.

And now a lovely evening. As I was putting the cows out in the night pasture it was getting dark, the half moon was bright overhead, and Venus was luminous in the purple afterglow. A light mist was settling in the vales. I lingered by the pasture gate

From the darkness of the woodlot far up the hill came a screeching sound, like barbed wire pulled too tautly through a staple — two young raccoons squabbling. The peepers sang in the marsh. A woodcock in the alder bushes near the gate began calling: peent, peent, peent, and then flew up high with wings twinkling like a bat, and his song twittering like a skylark in ecstasy. He was doing his 'sky dance' to attract his mate.

With all this, who needs TV!

SUMMER

FRED'S ULTRA-LIGHT

Look! Up in the sky! It's a bird. It's a plane. It's... a flying chain saw! Aw, shucks, it's only Fred Nagele, the flying farmer in the airplane he put together from a mail order kit.

You are likely to see Fred up in the air any day or evening when his farm work has slackened off. I saw him the first time just at sunset when I was up in the high meadow hayfield gathering some loose bales. First I heard a distinctive rattle like a faraway outboard motor. I looked up and about a thousand feet overhead saw what looked like a hang glider — a long span of wing covered with red sail-cloth. I could just make out Fred, slung in a harness seat beneath the contraption. It looked about as substantial as the seat of my daughter's swing dangling from the branch of the apple tree in our back yard.

Fred was out for a half-hour aerial tour of the farm hills on the north side of the Mohawk River. I envied him his bird's-eye view on such a serene evening. I watched until his craft was a mere speck silhouetted against a bank of rose pink clouds in the sunset sky.

Fred has been interested in airplanes for a long time. Way back in 1960 he had a student license and bought a Piper Cub for one thousand dollars. He also had a girlfriend. Within a few years he had to make a choice — keep the airplane and give up the girlfriend, or sell the airplane and get married. He couldn't afford both.

"I got married," he said.

Twenty years later, and now the father of a large family, he saw an ad in a flying magazine for a 'Weedhopper'. It was a do-it-yourself airplane kit — an ultra-light. The kit cost two thousand one hundred and ninety-five dollars. He sent away to Ogden, Utah for it.

During the winter of 1980-81 he assembled the ultra-light; it took him fifty hours, working spare time between farm chores.

His first flight was on July 4, 1981 from Nellis airfield. Nellis is a farm country airstrip two miles from Fred's farm on the south side of the Mohawk River. It has one blacktop runway.

Fred took the plane a thousand feet up that first flight. That is about the altitude of most of his flights. "At that height ground features still have plenty of detail," he says.

Fred's 'Weedhopper' has a cruising speed of thirty-five miles an hour. To me, that seems ideal. It must give you the feeling of being like a hawk, gliding ever so lazily up in the sky.

"The plane has one fault," says Fred. "On a hot muggy day it won't fly." He explains that hot, moist summer air won't give 'lift' to the wing.

Technically, Fred's 'Weedhopper' is classed by the FAA as an ultra-light because it weighs under two hundred pounds. Ultra-light aircraft are not subject to regulation by the FAA. Anybody can buy a kit, put one together and fly.

During the summer Fred has his plane up in the air almost every evening if the weather is fair. When he is still a long way off I hear him coming over the hill, up from the river valley. It takes him about six minutes to cover the four miles. He comes here to fly over his dry cow pasture to see if any of the cows have freshened.

"You can spot a calf real easy from the air," he says. "Even if it's bedded down in tall grass. Sure saves a lot of walking ."

* * * * * *

Fred's flying in the ultra-light isn't limited to summer. He is often out in winter. I remember one February day in particular; it was during a cold snap that lasted more than a week. The thermometer dropped to twenty below every night, and never got above zero in the daytime. Late in the afternoon when I walked to the barn to start chores, I heard his plane approaching. I looked up and saw the broad red wings. It made me shiver to see him dangling underneath, strapped in his chair.

I watched as his ultra-light circled and then slowly headed back toward the river valley. As the plane got further away it looked remarkably like an eagle carrying a baby in its claws, carrying it back to the nest.

It's a wonder he doesn't freeze to death up there, with the temperature zero and dropping. But he says it's no colder flying at an altitude of one thousand feet, put-putting along at thirty-five miles an hour, than it is snow-mobiling on the ground and doing sixty — if you're dressed for it, that is.

Fred wears four jackets, a sheepskin cap with the ear lappers tied under his chin, goggles, heavy mittens, and insulated boots and claims that keeps him warm. Even so, he doesn't stay up long — half an hour is the limit.

"The first ten or fifteen minutes isn't bad," he says. "But then the cold gets to you. It lets you know where the tiniest holes are in your clothes."

Fred gets a bird's eye view of all the countryside farms in winter. "One farmer still has heifers outside," he says. "They are being fed big round bales on the snow."

Mostly, Fred is just curious to know how many farms are still able to spread manure in this deep snow.

"Most of them are still spreading," he says. "But awfully close to the barn. There's nobody taking the tractor and manure spreader out to the back fields!"

TIFFY'S FOAL

On the last day of June we baled five loads of hay in the Hickory Lot, finishing up just before afternoon milking at five o'clock. As soon as we put the cows out to pasture at six o'clock, Randy and Larry, two high school boys from the village who are summer workers here on the farm, helped me unload the last two loads at the elevator and stack the bales in the hayloft.

My young daughter, Ann, rode with me when I drove the boys home at eight-thirty. On the way back from the village we stopped at Stewart's for ice cream cones. By the time we got back to the farm it was nine o'clock and dark.

We had let our horses graze on the lawn for an hour at supper time. My daughter and I went out to put them back in the pasture for the night. We have two Arabians. Our brown mare is named Tiffany; she was due sometime soon with her first foal. Our grey stallion is named Allegro; he is fifteen years old and gentle as a kitten.

The horses heard the grain rattling in the pail and walked across the lawn to us. Their shapes looked ghostly in the darkness.

Tiffy pressed her velvet nose to us; she breathed a long sigh as we petted her. "Soon now," said Ann. "Soon you'll be a mother."

The horses followed us across the lawn to their shed in the pasture. We dumped the grain in the manger and closed the gate for the night.

The next morning was the first day of July. When I went out to bring grain to the two horses, they were nowhere in sight. The horse pasture is divided into two sections; the sections are separated by a four-acre patch of brush, mostly alders. I rang the bell which usually brings the horses running from wherever they are in the pasture.

I waited a few minutes, rang the bell again and called. "Tiffy! Allegro!"

Still no response

I called again. This time I heard the stallion whinny from the far corner of the hill. I rang the bell and waited again.

Presently Allegro came trotting down the path through the brush, his neck arched proudly and his white tail flying like a banner. Tiffy would be a short ways behind. She usually came at a slower pace, being heavy with foal.

But this morning Allegro did not prance up to the shed and dip his nose in the grain bucket. He spun around on the grass and thundered back up the path.

"What in the world?" I thought.

I rang the bell vigorously and called. I could hear him trot back down the path. He emerged from the bushes and looked at me; he bobbed his head up and down, then whirled around and again cantered away through the bushes.

"I do believe he's trying to tell me something!" All I could think of was that some mishap had occurred. Tiffy must have injured herself!

I climbed over the fence and hastily started up the path. Thoughts flooded my mind. Tiffy had gotten her head caught in the crotch of a tree. Tiffy had slipped in a hole and broken a leg.

When the path emerged from the bushes, I skirted the back pasture toward higher ground to get a better view of the entire area.

"Tiffy! Tiffy!" I called as I walked.

No response.

I climbed higher on the hill and craned my neck, my eyes searching the pasture. Suddenly I stopped.

In the center of the field the pasture dipped in a bowl-like depression. There stood Tiffy. She turned her head and looked in my direction. And there by her side was a little brown foal.

I walked down through the dewy grass to them. And it was a filly at that!

A short distance away, Allegro trotted and pranced proudly

On long wobbly legs the foal moved closer to the mare and reached its nose for the teat. It never ceases to amaze me — how do they know where to look?

A few paces away in the matted down grass I found the afterbirth. All was well.

I hurried home to wake my daughter and tell her the good news. Together we hiked back up the path, carrying a bucket of grain for the mare.

The filly was unafraid and allowed herself to be hugged and petted. After an hour, when she was steadier on her feet, we walked the mare and foal down to the lower pasture and the horse shed.

Allegro cavorted in the lead, whinnying, turning back to encourage us.

"The filly looks just like her mother," Ann said. "She even has a white diamond on her forehead. Dad, why don't we name her 'Diamond'?

So we did.

JUNE HAYING

Our haying equipment was ready to roll on the first day of June. Unfortunately, the weather wasn't cooperating.

May had been exceptionally hot and dry, but by Memorial Day the weather turned chilly; the first days of June were overcast and cool. Every morning sent down a light sprinkle of rain, just enough to dampen the blacktop on the county road; turbulent clouds went scudding overhead; the sun shone fitfully or not at all.

It was the kind of weather the forecaster called unsettled. It certainly wasn't haying weather.

Since we were all primed for work, and getting edgy by being idle, we began to tackle all the jobs we had 'saved for a rainy day.'

Patrick, the hired man, got the claw hammer and pry bar and took apart the old icehouse. It hadn't seen an ice harvest for at least thirty years; lately it had become just a storage shed. During the last heavy, wet snowfall of April, the interior timbers collapsed, and the shed roof settled gently to the ground, taking the sidewalls with it.

The icehouse was ancient, the boards venerable. The planks and four-by-fours still were sound; but the boards were not good enough to use, and too good to throw away. He stacked the boards on a pile; they would come in handy some day; they always do on a farm.

While Patrick was dismantling the icehouse, I stretched some new fence wire along the upper heifer pasture. Then I mended the wooden manger in the lower heifer pasture.

The next day the weather continued to be unsettled, so we began sawing up the big dead elm trunk and splitting the blocks for firewood. Another day's work with the log-splitter netted us four cords of well-seasoned wood.

The fifth day dawned cool again, with a misty shower at mid-morning. We puttered around the barn, washing windows and nailing loose clapboards. With a vengeance, I got the ladder out and climbed up to clean a starling's nest out of the eave corner of the roof, then cut a piece of trim board and patched the hole.

Starlings can be a nuisance because they are such aggressive birds. Since they are cavity nesters, the least knothole or crack in a clapboard attracts them; they chip away at the wood with their beaks until they enlarge an entrance hole. Then what a mess they make, carrying in hay and twigs to build a nest, and spattering the barn siding with bird lime. Even the bluebird nesting boxes aren't safe from their depredations; they harass bluebirds, evict them, and usurp he nesting boxes.

Who was that misguided man of the early 1900's whose goal was to bring to America all the birds mentioned in Shakespeare's plays? He imported starlings and English sparrows and set them free in New York City's Central Park; now they overrun the entire country. A plague on him!

The sixth day of June was another damp day. We spent it repairing gates — gates inside the barn and gates outside in the barnyard. There are four gates in the barnyard: two in the holding area and two in the feeding area. "Too many gates!" complained Patrick.

A week of this uncertain weather was fraying our nerves. It's hard to believe but we were running out of rainy day chores.

What I should have done was chucked everything and gone fishing. I don't know why I didn't; it was perfect weather for it.

The dairy cows were enjoying this weather. It was ideal pasture weather: cool and damp and comfortable for the cows. The pasture mix of ladino clover and orchard grass loved the weather too and showed it with quick regrowth.

From what I read in the farm magazines, this was what the weather was like in New Zealand. I could take to it easily: green grass as far as the eye could see; frequent rain showers to keep the grass lush; cows on pasture the year round. And no need to make hay.

It was the pressure of hay-making that was making us antsy. With every day of cool, showery weather, the hay was growing more mature; quality would go down rapidly after the middle of June when the plants began replacing sugary sap with fiber as they prepared to flower and go to seed.

But there was nothing we could do until the weather itself straightened around. You can't make hay until the sun shines. We were just marking time with all our rainy day chores.

Early evening on he sixth day as I was bringing the tractor down from the heifer pasture, I saw a flash of blue in one of the wild apple trees by the stonewall. I stopped the tractor and stared; yes, it was a bluebird! And not just one, but a pair.

These were the first bluebirds I had seen in at least four years. In one of the apple trees was a dead branch with a hole in it drilled by woodpeckers. It would make an ideal home for the bluebirds, if the starlings didn't beat them to it.

I watched the bluebirds for about ten minutes. It didn't look like they were merely passing through They flitted among the branches of the apple trees as though they intended to stay awhile.

It seemed like a good omen. And to top things off, the sun went down in a bright red evening sky. 'Red sun at night, shepherd's delight' meant fair weather tomorrow.

It proved to be true. Next morning the wind was out of the west, the sky was blue and the clouds were scattered. I could feel the high pressure moving in.

Whistling as I went, I hooked the tractor up to the windrower. Then I drove to the Hickory Lot and mowed down half the alfalfa there. Ah, the smell of new-mown hay! And the warm breeze on my cheek! At last we were in the business of making hay again and it surely felt good.

CHLORINE SPILL

In the screened-in side porch comfort of his home on Klock Hill, Seeber Crouse was watching the afternoon baseball game on TV. He relaxed in a lounge chair, his hands clasped behind his head; his eyes left the TV from time to time and he gazed at his fondest view—his manicured back yard.

The fresh-mown lawn sloped to the creek bank where a row of eight-foot-high spruce trees bordered the water. The creek flowed in a long stretch of riffles before plunging over a waterfall; then it was lost in the wooded grove leading to the village of St. Johnsville.

There was a haze setting in over the waterfall. "Sure is getting dark early for June," Seeber thought, and glanced at his watch. "Only five o'clock. Maybe it's setting in to rain."

A short time later his neighbor further down Klock Hill stepped outside to weed his gladiola bed. He also noticed the haze drifting over the creek, but it seemed more like a wet fog to him. And there was something strange — the fog gave off a rank odor of clorox and made his eyes burn.

Long after dark, and a mile uphill in the country, Karen and John were watching the late movie on TV. They noticed an unusual amount of traffic on our normally quiet country road. Triumpho Road is a narrow gravel road only a mile in length; it is a crossover between two blacktop county roads — Mill Road on the west and Vedder's Corner Road on the east.

During the daylight hours the only vehicles seen on Triumpho Road are from the four homes located here: the pickup truck from Harold's farm; the two cars belonging to the Abeling family; the two cars of the Olendorf family, and my pickup truck.

To say that our country road is peaceful and little traveled is an understatement. Gretchen Abeling remarked at the end of one unusually busy summer day: "Gosh, there was a lot of traffic on our road today. I counted seven cars!"

So when Karen and John late that night heard tires crunching over the gravel road at the rate of one car every five minutes, they thought something was odd. Then Karen remembered it was graduation time at the local high school. "Must be kids out partying," she said, and smiled.

The next morning as I drove down to the village, I noticed that many of the trees overhanging the creek road had patches of withered brown foliage, as though they had been seared by fire or frost.

Then, from the man at the gas station I learned that there had been some excitement the previous night. St. Johnsville had experienced a chlorine spill. The fog Seeber Crouse noted actually was chlorine gas drifting above the creek. It came from several leaking chlorine cylinders in the village water treatment plant. The plant is located at the reservoir bordering Zimmerman Creek, a mile downstream from our farm.

The leak finally was discovered early that evening. Village water officials were unable to shut off the valves of the chlorine cylinders; the valves had become corroded by the chlorine and would not seal shut. There was only one recourse — to heave the leaking cylinders into the creek. This they did after receiving permission via phone from the Department of Environmental Conservation.

Immediately another problem presented itself. Zimmerman Creek flows through the west end of the village. Chlorine gas escaping from the creek could create a health hazard.

The village police car was dispatched to cruise up and down the streets, using a loudspeaker to warn residents to evacuate that portion of the village until the chlorine fumes dissipated.

People piled in their cars and abandoned the village. They spent all night driving back and forth on the country roads.

A couple of days later I hiked down through the woods and waded the creek below the reservoir. I wanted to see for myself the environmental damage inflicted by the chlorine gas. The creek in that area flows through a wooded ravine. The steep banks of the ravine acted like a funnel, channeling the chlorine gas. All the trees overhanging the creek had brown foliage, much the same effect a hard frost would leave. Ferns, weeds and wildflowers on the creek bank also were burned brown. The only trees that didn't seem to be affected were white cedar; they were conspicuous by their greenness.

In the creek itself I could see no fish — no chubs or minnows or crayfish. But if the fish had been killed, why didn't I find dead fish floating in the shallows?

All in all, the village was pretty lucky. Most of the vegetation eventually would recover.

Seeber Crouse was upset though about what the chlorine gas had done to his lawn shrubbery along the creek bank.

"Look at my spruce trees," he lamented. "The tops are all brown. They probably are going to die."

They were a sorry sight. Each bush and tree was faded to a sickly yellow-green as though it had been bleached.

"But don't worry," he growled, shaking his fist in the direction of the village. "They'll pay for it! They'll pay!"

FRED GOOMSBY

Fred Goomsby had a fine baritone voice of considerable range and power, and he liked to sing while he worked. Since a portion of my farm land bordered on his, I often heard a faint song echoing in the hills. But he never would be asked to audition for the Metropolitan Opera. The flaw, now that I had the leisure to analyze his singing critically, was the occasional note off-key; it was especially jarring because it came the moment when you least expected to hear it.

My discovery of this imperfection in his vocal powers came about quite by accident; it wasn't as if I was deliberately spying on him. I had walked up to a far heifer pasture to bring my youngstock some grain and see how they were doing. Fred happened to be cutting alfalfa; his hayfield was on the other side of the stonewall. I caught a glimpse of his red tractor through the bushes along the wall and saw that he was cutting the first swath around the field.

While my heifers crowded around the grain box savoring their treat, I sat on the stonewall and watched them. Fred's voice wafted up on the fresh morning air.

"One dream in my heart; one love to be living for; one love to be living for; it nearly was mine..."

There it was. And I flinched as the note came out sharp instead of flat. The heifers hadn't noticed.

It was a big hayfield, and by the time Fred had completed the first circuit and started coming round the second time, he was singing a different verse.

"To spend one night with you, in our old rendezvous, and reminisce with you, that's my desire..."

I tensed, anticipating the sour note, but it didn't come.

On his next circuit of the field, Fred was singing yet another melody. I had to give him credit; he had a large repertoire.

"Three coins in the fountain, each one seeking happiness, just one wish can be granted, which one will the fountain bless? Which one will the fountain bless? Which one will the fountain blesss...?"

I flinched as the last 'bless' came out in a slight screech. He had been doing admirably until that last note.

Presently he turned the far corner with the haybine and came clattering back. Above the racket of the sicklebar came a new song:

"For you, for you were meant for someone else..."

His delivery came through without a hitch. Maybe his pitch improves as he warms up, I thought to myself.

I was still sitting there on the stonewall when Fred swung around on yet another swath.

"Give me five minutes more, only five minutes more, let me stay, let me stay in your arms..."

The rich baritone voice came rolling up the hill. It was flawless. Even the heifers lifted up their heads to listen.

The next time around it was a new song:

"Six lessons from Madame La Zonga; she'll teach you the rhumba, the samba and conga..."

All at once something clicked in my mind. Fred was counting the number of swaths he was cutting! ONE dream in my heart. TO spend one night with you. THREE coins in the fountain. FOR you were meant for someone else. Give me FIVE minutes more. SIX lessons from Madame La Zonga...!

Fascinated, I continued sitting there on the stonewall. I just had to hear what other tunes he chose for numbering his trips around the hayfield. For instance, what in the world would he choose for the seventh trip now coming up?

I didn't have long to wait. Soon he turned the far corner and headed back. I strained my ears to listen and was rewarded by the faint bars of "Gonna take a sentimental journey, gonna set my heart at ease. Gonna take a sentimental journey, to renew old memories..." As the tractor chugged closer, Fred burst out the next line with a bellow:

"SEVEN! That's the time we leave, at SEVEN! I"ll be waiting up for HEAVEN, counting every mile of railroad track that takes me back..."

Spellbound, I waited for his next circuit. What tune would he dredge up for number eight? Wait — here it comes!

"I'll be down to get you in a taxi honey, so be ready 'bout a half past EIGHT! Now honey, don't be late..."

Why that son-of-a-gun! I laughed out loud. The heifers all turned to look at me.

"Did you hear that, heifers? That Fred is pretty cool!" The heifers looked at me curiously. "But Fred Goomsby will never think of a song with a 'nine' in it."

By now I had abandoned my seat on the stones; I was standing up and looking over the bushes expectantly at the remote figure on the tractor. The heifers lined up along the wall and peered over it.

Fred's voice rippled up to us on waves of song: "One little, two little, three little Indians, four little, five little, six little Indians, seven little, eight little, NINE little Indians..."

I slapped my thighs and almost whooped in delight, startling the heifers. "I never thought of that song!"

Soon the tenth swath was slithering down behind the haybine, serenaded by: "Give me TEN men who are stout hearted men, and I'll soon give you TEN thousand more!"

Fred had done it again. Well, the rest would be anti-climax now. I turned to go. I knew for certain there was positively no song with 'eleven' in it, or 'twelve' or 'thirteen' for that matter. Fred would have to start all over again with number one.

And he did. Only it was a different 'one.'

"ONCE there were green fields kissed by the sun; ONCE there were valleys where rivers used to run..."

As I walked away, still chuckling inside, Fred was rounding the corner again with "Tea for TWO, and TWO for tea, me for you, and you for me..." wafting on the freshening breeze.

At the far end of the heifer pasture I let myself out through the gate. The serenade was still coming, though fainter now, and I had to strain my ears to catch it:

"We're THREE caballeros, THREE gay caballeros, they say we are birds of a feather..."

I was reluctant to leave such a fine performance. It was a pity for that melodious baritone to be wasted on the earth and sky.

However, Fred Goomsby still had an audience that appreciated good singing. They were standing by the stonewall listening. The heifers, all twelve of them.

CATTLE CROSSING

My cows were going out to pasture after the morning milking. Since they have to cross the county road to get to the day pasture, I stood there to hold up traffic. This is a mere figure of speech because anything more than five cars an hour on our rural road is considered traffic congestion.

However, the unexpected always can happen. I was reminded vividly of that fact one summer evening when the cows were crossing the road on their way to the night pasture. It was just before dusk and the cows were ambling across the road in their usual unhurried manner when, in the distance, I heard a strange, wailing sound. Several seconds elapsed before I realized it was a siren.

The wailing gained in volume so rapidly I realized that whatever the vehicle, it momentarily would be rounding the far curve in the road and bearing down on my cows. Quickly I rushed the stragglers across, and none too soon.

A car whizzed around the curve and covered the eighth of a mile straightaway in what seemed three seconds flat. It was a small blue sports car and it traveled, as the saying goes, like a blue streak. As it whooshed past, I saw that it had on the rear deck a device popularly known as a 'spoiler', to stabilize the car at high speeds.

I caught my breath, because even a spoiler couldn't hold that car to the road on the next curve through the woods — the road took almost an 'S' bend at that point where it entered the woods; and the car must have been doing eighty miles and hour. I braced myself for the sound of the crash which inevitably must come, but there was none. He had made it.

What I did hear in the next second was the siren as another vehicle rounded the far curve — it was the sheriff's car in hot pursuit, although a full quarter of a mile behind. It whizzed past, tires squealing, engine roaring to deliver the last ounce of horse-power as it bore down on the S curve.

One second, two seconds — then it came — CRASH! The sheriff's car plowed through the woods and came to an abrupt halt against a pine tree.

But there was no such excitement on this hot July morning. The only car in sight was an old green Pontiac sedan coming around the curve not much speedier than a fast walk; it slowed to a stop as my last few cows plodded across the road.

I recognized the car. It was Fred Perry who always drove to the village about this time of day.

I stood by the car, waiting for the last cow to cross. As was her habit, she had to stop and itch her head on the concrete edge of the bridge parapet.

As I stood talking to Fred, I saw on the front seat beside him a small cardboard box full of red, ripe tomatoes.

"Where did you get those nice looking tomatoes?" I asked, with envy in my voice. "Have you been to the store already? Are those New Jersey tomatoes?"

"Naw," he drawled. "These are from my own garden. Picked them this morning."

"You did!" I said incredulously. "You've got ripe tomatoes in your garden, and it's barely the Fourth of July. I never get my tomatoes to ripen until the middle of August. How do you do it?"

"I start them in the house at the end of February."

"But even so, you can't put them outside until the frost is past, and that's Decoration Day."

"Naw. I put them out the first week of May and cover them at night if there's a danger of frost."

I couldn't take my eyes off those ripe tomatoes in the cardboard box on the car seat. The first red tomatoes of the summer are always mouth-watering.

"Boy, how I could go for some vine-ripened tomatoes today," I said.

"I'm taking these down to the Corner Market in the village," Fred replied. "They're always glad to buy these early local tomatoes." Then, seeing my hungry look, he reached in the box and picked out two gigantic tomatoes that were somewhat misshapen. He handed them out the window. "Here, take these."

"Gosh, thanks!" I said. "How much are they?" I reached into the pocket of my jeans.

"Naw, keep your money," he said. "I don't need any money. I've got enough money."

As I took the tomatoes and thanked him, he added, "What you can give me is your good left leg."

I stared at him quizzically for a moment, then realized what he meant. Fred had a 'game' leg. In his younger days he had been a logger. An accident in the woods had crushed his left leg. An endless round of surgery had restored its use, so he could walk with a limp, but it always was painful.

His logging days were over, but a woodsman who has been active right up through the prime of life can't tolerate sitting around the house. So he worked a large garden and the activity, which kept his mind off his aching leg, produced these early red tomatoes.

The last cow crossed the road. Fred shifted into gear and drove off. I walked over to shut the gate, conscious of the fact that I had two good legs.

HAYFIELD BIRD SONG

At the center of the twenty acre field the last waving stalks of alfalfa topple and spew out of the rear of the machine. Steering the tractor toward the perimeter of the meadow, I mow the backswath. Finished!

Halting in the dappled shade of a tall elm by the stonewall, I cut the engine. The growling roar of the discbine slows to a whine, then to a hum, then stops altogether. I flip off my sound-protector ear muffs. Blessed silence!

Stepping down from the tractor, I sprawl on the ground and lean against the elm trunk, relishing a moment of utter relaxation from the ongoing rush of haying season.

The smell of new-mown hay on the warm breeze and the serenity of the landscape basking in the noonday sun are soothing indeed. How peaceful and quiet, with only the soft cooing of a pair of mourning doves in the elm branches high overhead, and the cawing of a solitary crow, calling the rest of his flock to a smorgasbord in the new-mown hayfield.

The sound of the doves brings to mind those famous lines of Tennyson about the cooing of doves in "immemorial elms". The birds to which he referred were probably the English wood pigeons, but the sound is similar to our mourning dove, also known as the 'turtledove'.

It's fascinating how often the dove is referred to in literature. For instance, in the Song of Solomon, spring arrives when "the voice of the turtle is heard throughout the land."

Even the crow, cawing in the middle of my hayfield, has his day in print. The genus Corvus includes, besides our crow, the raven. Who can forget Edgar Allen Poe's raven quoting "Nevermore"? Poe admitted that, when he was conceiving the poem, he was undecided which talking bird to use, a parrot or a raven. He finally concluded that a parrot is comic, but a raven is sinister. And since the mood of his poem is meant to be dark and sinister, the raven won out.

Ravens are also prominent in Eurasian folklore, American Indian folklore, and even figure in the bible. A raven fed Elijah. And what bird did Noah send out from the ark? A raven.

Perhaps no bird has been celebrated so often in literature, notably in poetry, as the nightingale. Wordsworth, Swinburne, Matthew Arnold, all paid tribute to this songster of the midnight hours. In his "Ode to a Nightingale", John Keats speculated if it was "perhaps that selfsame song that found a path through the sad heart of Ruth when, sick for home, she stood in tears among the alien corn."

We don't have the nightingale here in America, but we do have an equally persistent songster — the mockingbird. In his poem, "Out of the Cradle Endlessly Rocking", Walt Whitman immortalized this bird. He tells of a pair of mockingbirds nesting on Long island; one day the she-bird is not on the nest, and is seen no more. But the he-bird keeps singing faithfully all summer long.

The crow in the middle of my hayfield has finally coaxed the rest of the flock to join him. A dozen black birds walk the swaths, looking for nests of field mice.

In the air above the meadow the barn swallows glide and soar. Yes, even the swallows have found their way into verse, thanks to Robert Browning: "And after April, when May follows, and the chaffinch builds, and all the swallows!"

Then there's the song, "When the swallows return to Capistrano." Why don't we hear when swallows leave Capistrano? Perhaps because leaving marks a sad time, the end of summer.

The swallows swoop and glide and soar high over the meadow. They are so graceful, I could watch them endlessly. Beyond the meadow, the swallows glide over the pond in the heifer pasture, dimpling the surface as they drink on the wing.

At the pond a solitary pair of Canada geese are nesting. They came with a larger flock migrating north in March, but for some reason chose to remain here for the summer rather than follow the others further north. This is the bird Thoreau called a true cosmopolite since on its return migration in the fall "it breaks its fast in Canada, has lunch in Ohio, and plumes himself for the night in a Louisiana bayou." (In this respect, all migratory birds are cosmopolitans.)

The honk of the Canada goose can't be considered a song. But there is no more majestic sight or sound than a phalanx of Canada geese winging northward in spring, their clarion calls drifting down like "goose music" to us earthbound mortals.

The most haunting sound here on our northern lakes is the cry of the loon. On a dark night, the laughing maniacal songs of a pair of loons is eerie. The melody echoes from shore to shore in pulsating rhythm, like flow of the northern lights in the night sky.

But enough of this reverie. I must up and away. Haying waits for no man!

THE CAMPERS

I couldn't help myself. I rolled over in bed and laughed out loud. Outside the bedroom window the black night was lit up by fearful flashes of lightning. Thunder crackled and rolled. Rain poured down in torrents.

All I could think of was those five people camping on top of the mountain at Nine-Cornered Pond. I could picture them huddled under the scrap of canvas they were using for a tent, a canvas they borrowed from my hay baler.

Surely it must be raining there too. Although the mountain was twenty miles to the north, this deluge seemed more than an isolated thunderstorm. Shortly after nightfall the heat lightning had begun flickering around the horizon. Along toward midnight thunder rumbled ominously, the wind increased, and then the heavens opened up.

And all this after the hottest and driest August in years. In fact, it had been so continuously hot and dry that when the campers left on their overnight excursion, they almost didn't bother to bring the canvas.

"It's just extra weight," Jimmy said. My nephew, Jimmy, was ten years old and on a summer visit to the farm. He and cousins Mike, aged nine, and Sue, aged seven, had persuaded my older brother, Joe, and his wife, Betty, to take them on a real camping trip.

After two weeks of pestering, Uncle Joe gave in. That afternoon they assembled their sleeping bags and cooking gear and stopped at the barn to borrow a canvas for a tent.

The kids were raring to go. They had a duffel packed with the makings of pancakes for breakfast — a small box of flour, two eggs carefully wrapped and cushioned — along with a griddle to use for cooking on the campfire.

As I did the evening milking, the sweat dripping off me in the hot barn, I wished I were going along with them. It would be refreshingly cool up on the mountain. And there would be leisurely hours to fish for trout.

But now, with this midnight deluge, I was glad for the shelter of my house. It poured for at least an hour, then let up for five minutes, only to pour even harder.

At five o'clock it was all over. The sun rose and water dripped from the barn eaves as I began the morning milking.

It was not yet seven o'clock when the campers drove into

the farm yard and walked in the barn. A flock of chickens caught in the rain couldn't have looked more bedraggled. Their wet hair stuck to their heads, their damp clothes clung to their bodies, and they looked like they had all spent a sleepless night. But they still managed to smile as they told me the details.

The two boys had slept out in the open in their sleeping bags. Susan slept in the tent with her Aunt Betty and Uncle Joe. When the thunderstorm broke, the boys braved it for a short while before seeking refuge in the tent.

"I knew it must be raining hard when Mike and Jimmy came crawling in the tent," said Sue. "It would take a cloud burst to make those two come in out of the rain."

"How would you know?" teased Mike. "You had your head under the blanket because you were afraid of the lightning!"

"Golly, how the lightning did flash," added Betty, and shuddered at the remembrance, she who felt uncomfortable during a thunderstorm even indoors with all the windows and draperies closed. It must have been a rousing ordeal for her with the scant protection of that little scrap of canvas as a makeshift tent.

But the three kids were indomitable to the end. With the unquenchable enthusiasm of youth, they avowed they had a wonderful time.

"That Jimmy is a real outdoorsman," marveled Sue. "You should have seen how easy it was for him to start a fire this morning with all that wet wood!"

Jimmy tried to look nonchalant under the lavish praise but all the proudness of his ten years broke through and he grinned from ear to ear with pleasure.

"How were the pancakes?" I asked.

Five voices chimed in unison, "Delicious!"

They went out of the barn on the way home to change in to dry clothes. As they were leaving I heard the children clamoring: "When are we going out again, Uncle Joe? We've got to camp out for a whole weekend! When are we going?"

RURAL CEMETERY CENSUS

Allison Smith did not realize the extent of the task she had set out to accomplish. Back in the February doldrums she had been wracking her brain in search of a worthwhile spring project to plan for and look forward to beginning as soon as winter was over; one suited to her capacity as township historian. She told me that it seemed like a realistic goal: compile a written record of all the rural cemeteries in the Town of Arden.

The more she thought about it and talked it over with other members of the historical society, the more enthusiastic she became. No such data now existed. It would be a real contribution to the historical record of the township.

People had lived, died, and been buried here for more than two hundred years, ever since Emmanuel Arden first set foot in the area back in the 1780's. Yet no permanent record existed of what families were buried where. Several times in the past few years Allison had received letters from descend ants of former residents, letters asking for assistance in tracing family genealogies.

She always was glad to help in any way she could. Old newspapers, family Bibles, baptismal records — all these were valuable aids, but there were many gaps. The further one went back from the year 1800, the more difficult it was to find any written record.

That's why the rural cemeteries were so important. Paper records will molder, crumble, and turn into dust. But the tomb stones in cemeteries have their records engraved in stone.

Many times Allison had tracked down someone's obscure ancestor by visiting a remote cemetery in some farm pasture and reading the names and dates chiseled in granite — dates of birth, marriage and death.

But nothing is eternal, not even granite. Two centuries of rain, sleet and frost had taken their toll. Carvings on some of the older gravestones had eroded to such an extant as to be almost illegible. Gravestone rubbing was the only way to read those. A large sheet of blank newspaper stock was taped over the stone and then rubbed with a crayon to make the lettering stand out. Every time Allison did this it brought back memories of herself as a school child in the second grade, and how she used to put a scrap of paper over a penny coin and rub it with a lead pencil.

When the snow melted in April, Allison began her project in earnest. She already had compiled a list of the location of each and every rural cemetery. It was amazing how many there were in the township — twenty-nine. An early count had turned up only twenty, but as word of her project spread on the rural grapevine, several farmers stopped at her house to call attention to other cemeteries that were hidden away in a back field, or corner of an old woodlot.

Several of Allison's friends volunteered their assistance, and she was grateful. The project was becoming more involved than she anticipated. For one thing, many of the rural cemeteries could not be reached by car. Often, she had to park on a farm cow lane and walk half a mile through fields.

At the cemetery she first made a map — a ground plan — to show the location of each grave. Then she meticulously recorded the inscription from each headstone.

As May progressed into June, Allison began to lose her volunteers; first Sarah, then Hazel. They were sorry, they said, but they just could not spare the time from their big vegetable gardens or the weeds would get ahead of them.

The end of June approached with her project only half complete. Suddenly, and completely unexpectedly, Allison found herself in the hospital for surgery. After a week she returned home to convalesce; and then it was already July. She was dismayed — the completion date for her project was August 1st; but she was undaunted. She coaxed and pressed into service her two young daughters, Missy and Paula, ages ten and twelve.

Together they visited the next cemetery on her list. Allison found herself almost completely exhausted after the hike to it. She really ought not to exert herself so strenuously, especially so soon after returning from the hospital, but the project had to be completed in time for the Town Bicentennial.

    She sat down on one of the flat coping stones of the limestone wall surrounding the cemetery to catch her breath. Then she took out her notebook and pencil and wrote down the information as the two young girls read the inscriptions for her. One of the head stones was worn almost smooth and had to be rubbed. Missy and Paula both thought that was great fun. At the end of the afternoon, a strawberry sundae at the drive-in was their reward and the girls looked forward to the next day's jaunt.

    In this manner they continued for a week. If the cemetery was small, and all the headstones legible, they were able to compile information in a single visit. A large cemetery, or one with toppled headstones that had to be raised and washed off, required two or more afternoons of work.

    Allison was comforted that her two young daughters' enthusiasm did not diminish as the days went by. She had expected their interest to wane, but Missy and Paula actually became more engrossed, especially when they came across names related to those of a previous cemetery.

    One July afternoon they were in the Flanders cemetery. The day was so hot and humid that Allison suggested they leave, eat an early supper, and return at seven o'clock when the evening had cooled down. With daylight saving time, there would still be nearly two hours of daylight by which to read gravestones.

    This was agreed. Several hours later they were back; it was cooler and more comfortable working. Allison sat down on a low monument with her notebook, and the girls commenced reading inscriptions and calling the information to her.

As nine o'clock approached the evening shadows grew longer; the girls, who had been chattering and giggling near a tall granite monument, suddenly became silent.

"Did you hear that?" Missy asked her sister.

"Yes!" Paula answered, almost in a whisper.

"It sounded just like bones rattling!"

"Mother! We want to leave!" they said in unison. "All of the sudden it's spooky here."

"Nonsense!" Allison said. "You girls are too big to believe in ghosts. And besides, we only have two more gravestones and we'll be finished here."

The girls relented and retraced their steps back toward the tall granite marker. But they had been there not even a full minute when Missy stiffened and said, "There it is again!"

Both girls jumped and ran back to their mother.

"We're scared! We heard bones rattling again! We really did! There are ghosts in this cemetery! We want to go home!"

There was no reasoning with the silly girls. Ghosts indeed! But now it was getting dark. There was nothing to do but leave.

By next morning the girls had lost all interest in the graveyard census they once found so absorbing.

Allison was totally exasperated with them. She phoned her friend, Hazel, to explain what had happened. "Hazel," she pleaded, "couldn't you possibly help me again? I absolutely must finish the census and time is running out."

"Did you say you were in the Flanders cemetery last night?" Hazel asked.

"Yes, we were."

"As I recall, that cemetery is at a farm on the hill right next to Big Nose Mountain." There was a long pause. "That place is full of rattlesnakes. That whole side of Big Nose mountain is notorious for rattlesnakes. Didn't you know?"

Allison gasped. "You mean..."

"You can imagine how those gravestones hold the heat on a cool evening. Snakes love warm stones like that."

"My God!" Allison felt herself trembling. "Do you mean it was a rattlesnake the girls heard? My God! That was a close thing! Thank heaven Missy and Paula persuaded me to leave!"

"Those were bones the girls heard rattling all right," Hazel continued mildly. "But not the bones of a human skeleton."

SUMMER VACATIONS

The milk tester showed up nearly a week early in August and, as I proceeded with the milking, he explained the reason.

"My wife and I are going on vacation next weekend, so I want to get the herds caught up before we leave."

"Vacation, eh?" I said. "Boy, I wish I could get away. Where are you going?"

"Up to the St. Lawrence River. We're going to take the Thousand Islands boat ride, because my daughters will enjoy that. Then we're going to spend a couple of days across the river in

Canada; there's a restored colonial village we want to visit."

"It sounds like a nice relaxing vacation," I said with a touch of envy in my voice.

"That colonial village is a fascinating place," Gerald added. "We were there once before when our daughters were little. They're old enough now to really appreciate it."

While the pulsators hissed their soft refrain, Gerry went on to explain a little bit more about the village.

"All the people dress like they did two-hundred years ago. There's a grist mill where they grind flour, and there's also a colonial bakery. There's a blacksmith shop and a wheelwright. They have cows and sheep and draft horses; and they cut hay the old-fashioned way and put it in the barn loose. Just a leisurely walk through the village makes you feel like you are back in the 1700's."

"Sounds like an ideal vacation," I said.

"Then we want to spend one afternoon at the Eisenhower Locks on the seaway, and go through the power plant; they have a good tour."

After milking was finished and Gerry was carrying the milk meters and milk samples out to his car, he called out over his shoulder, "I'll be thinking about you sweating in the hayfield while we're cruising down the river!"

Half an hour later the milk tanker backed into the farm yard and Dave, the driver, came whistling in to the milkhouse.

"You sound happy," I said.

"That I am, that I am," he replied. "I'm taking five days of vacation, starting tomorrow."

"No kidding? Where are you off to?"

"I'm going deep sea fishing."

"Is that right!"

"Yup. There are four of us going — a friend of mine and his buddy, and my father-in-law."

"Where are you going fishing?"

"Off of Rhode Island. We've chartered a boat, and we're all set."

"Have you ever been deep sea fishing before?" I asked.

"Nope. But I've always wanted to go. My wife's father goes every year, and he finally talked me into going."

"What will you fish for?"

"Bluefish, cod, flounder — whatever's running."

"Boy, I love grilled bluefish," I said.

"We're going to live right on the boat for three days. Eat what we catch."

"If you have any extra bluefish, bring one back for me," I said. "My mouth is watering for a nice broiled fillet."

I was still thinking about broiled bluefish when the tanker pulled away, and I was left to wash the bulk tank.

After breakfast, the high school boy who works for me showed up for the day. I backed the tractor up to a wagon and Derek dropped in the hitch pin. We headed out to the hayfield just as a car pulled into the driveway. It was Fritz, the grain salesman.

"I won't hold you up but a minute," he said. "I just wanted to see if there was anything you needed, because I'll be gone for the next three weeks."

"You must be going on vacation, too," I said.

"Yes. My wife and the boys and I are going to the Bahamas."

"Ohhhh!" I said. "Pardon me if I drool."

Fritz smiled and went on to explain that his family had been planning this vacation for the past two years. "We're going to the outer islands which are undeveloped and not commercialized in the least."

"How do you get there?"

"We fly to Miami and then take a small commuter flight to the main island. From there we take a water taxi to the outer islands."

"If it's undeveloped, where do you stay?"

"We have rented a beach house owned by a congressman from Illinois. We're going with another couple and splitting expenses, so it really is quite reasonable."

"So it's a housekeeping cottage? Where do you buy your groceries if the island is undeveloped?"

"The island we'll be staying on is called Martha's Cay. It has a small village of about a thousand people, and there's a general store where we can buy groceries, and also a small restaurant if we get tired of our own cooking."

"I should think the Bahamas in August would be quite hot for a vacation."

"Not at all. The temperature never gets above the 80's, and there's always an ocean breeze."

"So you'll fish and swim and lie around on the beach."

"That's about the size of it. As I said, we've been saving for this vacation for a couple of years. Our oldest boy is fifteen now, and he'll be leaving home to go to college before you know it; so we thought we'd better take a family vacation and not put it off any longer."

"What kind of fishing is there in the Bahamas?"

"Close inshore there are bonefish and barracuda. Further

out there are wahoo and dolphin." It sure sounded like a dream vacation to me.

We chatted a few minutes more. I felt more and more envious as he talked about the wonderful vacation they were going to have. As Fritz got ready to leave he said, "If you need anything while I'm away, just call the feed mill."

"No," I said. "I'll just fly down to Martha's Cay instead!"

Later that afternoon Derek and I returned from the hay field; we were hot, dusty, sweaty, and plumb tuckered out.

The vet was there, checking one of our cows. As he put in the thermometer to take her temperature, he said, "I'll need a pail of warm water."

Derek went to fetch the water, and I held the cow's tail. Tom finished the examination and infused the cow. While he was washing up, he said, "This cow should be checked again in a few days. I'll be gone, but I'll leave word with one of the other vets at the clinic."

"Where are YOU off to?" I said.

"I'm taking the family to Cape Cod for a week's vacation."

I resisted the urge to take the pail of soapy water and dump it over his head.

"You're the fourth person to stop at this farm and tell me you're going on vacation! Do you need anybody to carry your luggage?"

THAT DARN BULL

When Fred Nagele's old red pickup pulled in my barn yard I knew right away what was the matter. "That darn bull!" I said to myself.

My heifer pasture was separated from Fred's heifer and dry cow pasture by one of my hayfields. All summer long my four yearling heifers and my two-year-old bull had minded their own business. But I baled hay in that field in July. Three weeks later I opened up the gates and allowed my heifers and bull to graze the afterfeed. That proved to be a mistake because now only a stonewall separated my animals from Fred's. And there was a gap in the stonewall where a gate had existed in bygone days. The gap was closed off by three strands of barbed wire but it remained a gap nevertheless, and a temptation to any animal with wanderlust.

My bull didn't go through the fence — not at first. One of Fred's heifers came in heat and crawled under the fence to get to the bull. The first I knew of it was when my neighbor on the hill, Karen Olendorf, who lived across from the heifer pasture phoned me.

"Did you put more heifers in with your bull?" she asked.

"No," I answered with misgivings. "Why?"

"Well, I just now looked out my kitchen window at your pasture and there are about a dozen animals around the water trough — even three cows."

"Cows?"

"Yes. One looks like a Brown Swiss. Another looks like a Jersey-Brown Swiss cross."

"Oh, no! Those are Fred's dry cows!"

And now here was Fred himself asking if I would lend a hand to separate the two groups of animals. One of his dry cows had freshened and he wanted to take the cow and calf out of the pasture.

I hopped in my truck and followed Fred up the hill. Using a pail of grain, I lured the bull far enough away for Fred to get a rope halter on his cow and lead her out the gate and on to his truck. Then he carried the calf toward the gate.

The calf bawled at the top of its lungs. The cow lifted her head over the truck rack and mooed. The bull gave an answering bellow; he whipped around and headed for the gate. Fred got the calf out and got the gate closed just in time.

The calf continued to struggle and bawl while Fred lifted it on the truck. The bull tossed his head high and his eyes gleamed fiercely as he bellowed again. For a few anxious moments I thought he was going to charge through the wire gate. The way he carried on you would have thought the calf belonged to him.

Just as his big head came close to the wire gate, and I was all set to warn Fred and the kids to run for cover, the bull's attention was diverted.

Fred's brindle heifer — the one that had been in heat — caught his eye. The bull turned and trotted after her with an air of 'This is where I came in.'

I let out my breath. "That was close!"

Fred smiled. "Beef is pretty high right now. He ought to fetch a good price at the auction."

"He'll be headed that way before long," I said. I retrieved the empty grain pail from the pasture. "I'll put my bull and heifers in my heifer barn tomorrow, Fred. Then you can drive your stock back and we can fix the fence."

Within the next two days, both groups of animals were in their proper pastures and the fence was repaired. An uneventful four weeks passed.

Then one afternoon as I was bringing the dairy herd from pasture for milking, a sporty blue car stopped and a comely brunette stepped out. I recognized her as Fred's younger sister, Irmgarde, who had recently married and moved to California. After the hellos, Irmgarde had a request to make.

"Dick, I hate to bother you, but do you think you could get your bull out of my brother's pasture? I want to sell my Brown Swiss cow before I return to California, but first the vet has to trim her feet and check her for pregnancy."

"Is that darn bull back there again?"

"I'm afraid so. I've been going to the pasture every day to bring grain to my cow and get her used to the halter again. Today when I went to the pasture I saw the bull in there and I backed out fast."

"I'll get him out," I promised. "Then I'll close off the gates to the hayfield so he can't get near your brother's pasture again."

"The vet is coming Friday afternoon," she added. "And I wondered, would it be too much to ask if I could bring my cow in to your heifer barn? It would be easier for the vet to work on her there than in the pasture. The first time the vet examined her was six months ago. He thought she lost her calf but he wasn't sure, so she needs to be rechecked."

"Of course," I said. "In fact, I'll bring her in the barn myself. It's the least I can do considering the bother the bull has caused."

The next day I walked to Fred's pasture with a pail of grain. The Brown Swiss cow was with a group lying down. She saw me at once, and when I rattled the grain in the pail she stood up, stretched, and walked straight across the field toward me.

Using fence pliers, I moved the bottom two strands of wire away from the gap in the stonewall. The Brown Swiss hesitated at first about coming under the top strand of wire, but after tasting the grain again, she ducked her back and came through.

I swung the bottom two strands back and closed the gap. Then I marched across the hayfield, the Brown Swiss following like a lamb. All she needed was a thick leather neck strap and a bell to make me feel like a Swiss farmer leading his herd down from an alpine pasture. I picked up a stick just in case she became too playful with her wide horns, but I didn't need it. In ten minutes we were in the heifer barn and I put her in one of the empty stalls.

I was there when the vet came, and I heard the good news. Heidi — for that was the cow's name — was two months pregnant. That meant she had aborted previously, six months ago, and then my bull had bred her.

"Well, I'm glad I can sell her as a bred cow," said Irmgarde. I would hate to beef her. I've had her since she was a calf, and her mother too. I showed her mother in 4-H."

The vet trimmed both hind feet while Irmgarde stood by Heidi's head, petting her and cooing to her. Heidi was now ready for her new home.

"Some people near Cherry Valley bought her," Irmgarde said. "Their farm is right along route 20. I know they will take good care of her. If you are ever down that way, would you see how she's doing?"

"Oh, I will," I said. "You might say I have a passing interest in her too. I'd like to see what a Holstein-Brown Swiss calf looks like, come April!"

ROOFTOP RASPBERRIES

I never dreamed that I would ever have to stand on a house rooftop to pick wild red raspberries, but I did one day during the first week in July.

We were finally done with the first cutting hay. We were fatigued, yes, overtired from the great harvest we ourselves had desired, and I needed some rest and relaxation.

Monday morning dawned with a beautiful soft rain that began about five o'clock. When I got out of bed for barn chores and heard the soft hissing sound of rain caressing the earth, I felt a deep inner contentment. Twice blessed was this rain, coming as it did after three solid weeks of hot, dry weather. The first cutting hay was in the barn, so let it rain! The parched hay stubble would soak up this rain, and the second cutting would shoot up. And I knew what I would do today — go fishing!

As I walked to the barn I almost could hear the grass drinking the rain. The maple trees stretched out their boughs in gratitude as water dripped from their leaves.

Beads of moisture glistened on the orange blooms of the Canada Lily and the Turk's Cap. Droplets of water clung to the white blossoms of the Meadow Rue and Queen Anne's Lace, making them sparkle like jewels.

By the time milking was over and the cows were let out to pasture, and I sat down to breakfast, the rain had let up to the barest of drizzles. It was more like a mist than a drizzle, but I knew it would be an all-day affair.

The creek already was on the rise. I hadn't been fishing since May. It was time the trout and I had an encounter.

With the luxury of unaccustomed leisure, I finished breakfast and contemplated where I would fish. Since I had the rest of the day, why not go way upstream, a mile past Gabby's farm?

Down from the cellar beam came my hip boots. Out of the closet came my battered fishing hat and rain jacket. Off the shelf came the trout rod.

A few minutes prodding with a shovel in the garden produced a handful of wriggling red worms. I got my creel, gathered up rod and reel, and headed for the truck. Then I had a second thought and returned to the garage for a berry pail and stuffed it in the creel. The red raspberries were ripe, and I knew there was a large patch of berry bushes on the overgrown end of Bell Road not far from the creek.

At the old wooden bridge near Bell Road I strung the line through the guides of my trout rod, then tied on a nylon leader with a small hook and split shot. On the hook went two plump worms. I waded into the swollen creek and with an easy gesture cast the line upstream and let it drift down.

The rain pattered lightly on the water and dripped from my hat on to my nose. It was a warm rain, now letting up, now increasing to a drizzle.

I made successive casts, expecting at any moment to feel the quick tug on the line from a submerged brown trout. Slowly I worked my way downstream, letting the line drift into every likely looking eddy and riffle. The little pool below each boulder received a cast, but not so much as a nibble did I get.

After the better part of two hours, I had to admit this wasn't my day for trout. Like all unsuccessful fishermen, I needed an excuse; mine was that the trout began feeding as soon as the creek began to rise in the wee hours of the morning. Now, bellies full, they weren't looking at any bait.

Well, no matter, that's what the berry pail was for. I climbed up the bank out of the creek and emerged from the bushes into an open pasture, to the wonderment of Gabby's cows. They lifted up their heads from grazing and stared at me as I disassembled my trout rod.

I stared back, spotted the bull at the far end of the pasture and retreated into the bushes. I hiked back up along the creek bank through the woods to the plank bridge where I started. Then I tramped through the underbrush on Bell Road in search of the berry patch.

I found the big patch of shoulder-high brambles crowding around the old abandoned farmhouse. Plump red raspberries glistened like rubies. I took my pail out of the creel and began picking.

By the time I had worked my way around one side of the house, past the overgrown peony bushes, my pail was half full. These red raspberries were not scrubby little ones like wild ones sometimes are; these were the size of thimbles, plump, juicy and dead ripe.

My hip boots were ideal for wading through the briars. At one point I startled a little yellow warbler from her nest; she bobbed up and down on the raspberry stems, wondering what sort of strange creature had invaded her sanctuary.

At the side of the house I peeked into an open window frame at the ruined interior. Plaster sagged from the walls; bare lath was exposed like skeleton ribs; rubble littered the faded linoleum on the floor. But the house frame itself was still sound, although vacant how long — fifty years? They really built these old farmhouses, with eight-by-eight beams and four-by-four studs in the timber frame, not the puny two by-fours that are used today.

I continued working my way through the brambles to the rear of the house. Here the back porch had collapsed and the rusted tin roof was settled down into the brambles. That's when I climbed up on the roof and finished filling my pail, accompanied by low rumbles of thunder and brief showers.

Soon the clouds began scudding away; by four o'clock the sun was breaking through. I tramped back through the woods to my truck parked by the plank bridge. I was at peace with the world. My creel was full, though not of trout.

At supper there were two warm raspberry pies, fragrant and juicy. With each savory forkful, I could see in my mind's eye the abandoned farmhouse in the overgrown clearing, forgotten by all but the raspberry vines, a nesting yellow warbler, and a berry picker.

ROBERT ROWLAND

The road I travel on my way to the farm supply store in the nearby village of Fort Plain goes past the historic Palatine Church. Surrounded by pastures in which black and white cattle graze, the church and a cluster of four houses stand by themselves on a by-passed portion of the old turnpike, close by Caroga Creek.

The church was built in 1759 by the Palatine Dutch. They arrived in the Mohawk valley eighty years after the Pilgrims came over on the Mayflower. One of their settlements sprang up near Fox's grist mill on Caroga Creek, a short distance up stream from its confluence with the Mohawk River. They quarried limestone and near the mill they built the church. It is said General George Washington once worshipped there.

At the two hundred year anniversary of the founding of the church the Palatine Society, whose members trace their lineage back to those early settlers, agreed to its restoration. For many years the church had been without an organ, the original organ having died of old age sometime beyond living memory.

So when Robert S. Rowland, a seventy-year-old church organ builder whose roots sprung from the Mohawk Valley, was looking for a way to do something for his old hometown, he decided to build a new organ for the church and donate it in memory of his parents.

Over a period of several years, sandwiched in between his regular work of church organ building and repair, he built the organ in his workshop in the Hudson Valley community of Ossining. He built it as a free-standing organ, patterned after the pipe organs that were built in the American Colonies prior to and immediately following the American Revolution by such men as David Tannenberg and the four generations of Dieffenbachs.

Finally in the spring of 1978 the organ was completed in his workshop. Robert then dismantled the organ and trucked it to the Mohawk Valley. During the summer, he and his wife, Mildred, assembled the organ inside the Palatine Church. At the same time they acted as curators, because the church is open to the public during the summer as a historical shrine.

Trying to assemble an organ while answering questions from tourists can be frustrating at times. Although he himself is a member of the Palatine Society, Robert has no qualms about poking fun at those members who take themselves too seriously.

I happened to be there one particularly hectic day when there were a lot of visitors. Robert became somewhat annoyed at one persistent tourist. You know the kind — an old biddy with a long nose in everybody's business.

She saw Robert puttering around at the organ and said to him, "Waall, uh, what're you doing here anyway? Whooo are you?" When he seemed too busy to reply, she added, "Our group are all Palatines. We belong to the Palatine Society!"

Of course, Robert and his wife do too, but he didn't tell her that.

She persisted. "Waall, what's your name?"

Robert told her his name.

She looked down her nose at him and asked, "Are you one of those famous Palatines?"

"No," Robert said. "I don't happen to be one of those famous Palatines."

"Waall," she said. "Where do you come from?"

By that time I could see Robert becoming irritated. He couldn't concentrate on putting together the organ. He said, "I'll tell you, Madam, where I come from. This is a true story. You won't believe this, but it's true.

"Way back in 1709 when Queen Anne was having trouble with those old Palatines — they came over from the Palatinates you know, settled in England, and she had to get them out of the country because they had pigs, and chickens, and goats, and ducks, and geese, and lice and everything else. She decided to ship them over here to pick up tar and, you know, pitch pine for the English navy.

"And they landed in Rhinebeck, and from Rhinebeck they came up the Schoharie and arrived here. And she sent all those Palatines over in ten boats you know."

"Oh, yes," the woman replied. "Yes, I know all that."

"Well," Robert said, "what you don't know is that after the queen got them all shipped , shifted them out of the country you see and shoved them off shore, she had her scouts look all around England to clean the country up. And they found a lot of these old vagabonds, and tramps, and thieves, and ne'er do-wells and everything. They got them all together and they found that they were all named Rowland.

"So they took these old Rowlands and put them all together with all their filth, and put them on another boat that they knew would never reach port — unseaworthy you know — — and shoved them off to follow behind the Palatines.

"Now, they weren't Palatines, they were Rowlands. So the Palatines kept looking back, and looking at each other, and they said 'Here come those awful Rowlands in that other boat.' Now they said to their children, 'When we land, don't you pay any attention to those old Rowlands because they've got more lice than we have.'

"Well, anyway, the Palatines finally landed here and they looked back — and no Rowlands! 'Hooray,' they said. 'They've all drowned, boat and all!'

"But the Rowlands fooled them, and in a few years they started to sprout up all over the country — and that's where I come from."

The old biddy's eyes narrowed in disbelief and she sputtered, "I-I-I-I think you're fooling me!"

"I'm not fooling you one bit, Madam," Robert said, keeping a straight face. "That's the God honest truth." He turned back to resume his work assembling the organ.

She said, "That's a true story?"

Robert kept his serious look. "That's a true story because I made it up myself."

Her face reddened. "Now I know you're fooling me!" She stamped her foot, turned and left in a huff.

AUTUMN

POWERLINE R.O.W.

The wonderful aroma of alfalfa filled the air, and I breathed deeply of that sweet smell as the freshly mown third cutting came spinning off the reels of the mower-conditioner. A blue summer sky up above with scattered white clouds drifting lazily, a pleasant breeze out of the west — it was a glorious day.

A squadron of barn swallows kept me company. In much the same way as dolphins follow a ship at sea, frolicking in the bow wave, the swallows kept pace with my tractor and mowing machine in our voyage through the waving alfalfa. Soaring and gliding above me, they fed on the insects stirred up by the sickle bar of the passing machine. I never cease to marvel at the gracefulness of their flight. With a seemingly effortless stroke of their streamlined wings they ascended thirty feet or more, then banked on the invisible wind and swooped low to the ground in long, curving arcs.

'On wings of song' is an apt description of their flight. They seemed to partake of as much joy in negotiating the breezes as dolphins do when sporting in the waves. Swallows even look like miniature dolphins — the same sleek, torpedo shaped body — and even the shape of head and beak resembles the dolphin.

Every motion of their flight signals their exuberant joy in living. An occasional twitter could be heard as they talked to one another while flying. It resembled the clicking sound that dolphins make when communicating to each other in the water. Someday we may discover a swallow 'language', much as Cousteau delved into the mystery of dolphin talk.

"If I could talk to the animals, talk to the animals, take an animal degree; talk to the animals, and they could talk to me." How did that song go? My spirits soared on wings with the swallows. What a glorious day it was!

From this high meadow I could see for miles, and all creation was beautiful. The distant green hills, checkered with woods and farms, stood out in remarkable detail. The air always was this clear when the high pressure was moving in, bringing fair weather. A perfect day to begin the third cutting of hay.

The alfalfa was lush and thick in this field. The twenty-acre meadow diagonally downhill was greening nicely too, and would be ready in another week. But who was that damn fool driving through the alfalfa!

I couldn't believe my eyes! A vehicle of some sort had just

nosed over the crest of a knoll in that lower alfalfa meadow, heading smack down the middle toward the gate in the nearer end. It was a pickup truck.

As it got closer I could see the orange and white paint job; it was the electric power company's truck. It came through the gate and left my range of vision as it headed up a steep hill toward the metal towers of the power line.

The alfalfa in that meadow was almost knee-high. Anyone with half a brain would know they wouldn't be thanked for riding over it and leaving a trail of flattened alfalfa. I shut off my tractor and set off on foot to intercept the culprit.

I had to climb over the stone wall and walk halfway across the heifer pasture before I saw the pickup parked near one of the towers. As I walked closer, I could see a man seated inside; he was writing on a clipboard.

"What are you up to?" I asked

He leaned out the window . "I'm taking inventory." he said.

"Inventory? Does that mean you are counting the power poles?"

"No. I'm counting trees" He tilted the clipboard so I could see the paper he was writing on.

"The choke-cherry and sumac are desirable trees," he said. "The ash and maple and elm are undesirable."

At first I didn't understand what he meant. Surely he had things backward. Ash and maple were desirable timber trees. Choke-cherry and sumac were the undesirable ones; they were not much taller than bushes.

But he told me he wasn't looking at them as timber trees. He was looking at them as marauders that were growing within the power line right-of-way. Now I understood his usage of the terms 'desirable' and 'undesirable.'" It certainly was a new way of looking at things.

"The ash and maple and elm are tall-growing and will have to be cut down before they interfere with the high-voltage line," he said. "The choke-cherry and sumac and other brushy types will never grow tall enough to interfere with the power line and can be left alone."

"Well, shades of Rachel Carson!" I said. "Did you by any chance read her book *Silent Spring*?

He nodded in the affirmative.

"Do you recall that in one chapter she bemoaned the indiscriminate use of herbicides along power line rights-of-way? She said it would be much wiser, ecologically, to encourage the growth of low-profile bushes and shrubs."

He nodded again.

"She was a prophet thirty years ahead of her time," I said.

I told him about the last time the power line maintenance crew had been through; it was eight or nine years ago. They sprayed the full width and length of the right-of-way with brush-killer. It left a one-hundred-foot wide swath of dead brown grass and foliage. We didn't dare pick blackberries under that line again. Not to mention the heifer that shriveled to skin and bones and died a few weeks later.

"Things have changed," he said. "We don't spray any more without a permit from the state environmental agency. And our spraying now is only spot spraying; there is no more blanket spraying."

That good news was so encouraging that I almost forgot the reason why I had left my mowing and walked across two fields to talk to him.

"Do you always drive across hayfields?" I asked.

His smile was apologetic.

"I saw what appeared to be an overgrown wheel track in the hay, so I followed it."

"We did haul wagons over that field, but that was last month when the hay had been cut. Now that the alfalfa has regrown almost knee-high anybody familiar with farm country knows you don't drive in and knock tall hay down."

He looked chagrined. "Sorry about that. Guess I'll have to return to my old motto. When in doubt, park the truck and walk!"

BILL KECK

On the narrow blacktop road approaching the village, I slowed my truck almost to a stop to avoid colliding with a car parked half-off the pavement.

As I continued on past the car, I did a double-take; I couldn't believe what I was seeing. A young woman was standing in front of the car; she was facing the lawn at the side of the road; and she was squinting through the eyepiece of a camera, taking a photo of a small creature on the front lawn of Bill Keck's house. She was snapping a picture of a chicken.

The chicken — a white leghorn hen — seemed totally unconcerned. It was not much more than an arm's length from the woman and, being up on the raised terrace of the lawn, it was on eye level with the camera. Yet it was completely oblivious to being the object of photographic zeal.

Indeed, at first glance the chicken looked like a lawn ornament; it was standing motionless on one leg, with the other leg tucked up, as chickens often do when they are resting. I had to look twice to make sure it wasn't one of those ceramic lawn decorations.

But of course, it wasn't. What would Bill Keck do with a lawn ornament? His flock of chickens were his only lawn ornaments, and the majority of them just now were employed busily scratching here and there in the grass. A few were on the porch, resting on the front steps. A rooster showing off his brilliant plumage was strutting near the hand pump of the well cover, but he was out of the range of vision of the girl, or she would have been eager to capture him on film too.

The compact car — a shiny import — had license plates that showed it to be from away, meaning downstate near New York City. Its occupant doubtless was one of those weekend vacationers who come upstate in autumn, seeking views of the colorful foliage, and a little local color in the boondocks.

She would have found a colorful subject in Bill Keck's hen house. It was on the rear lawn, behind his dwelling. The hen house was only big enough for one person to turn around in. It was built with scrap lumber; the sides were patched with sections of corrugated cardboard.

The small chicken yard attached to the hen house was even more photogenic. It was fenced in by rusted chicken wire. In a futile effort to keep the chickens from flying over the top, Bill had covered it with an old rose trellis, a broken section of a TV antenna, and a bent aluminum lawn chair.

The chickens obviously did not spend much time within the confines of their fence. They ranged freely, laying those good natural free-range eggs. Once, a tourist, seeing the sign on Bill's lawn with the crude lettering 'Free-Range Eggs' had stopped and asked for some of the free eggs. Bill had to explain to her that the eggs weren't free; they were a dollar a dozen.

Bill Keck was one of a vanishing breed. He was nearly seventy, but he seemed ageless. A bachelor who had been a life-long hired man, he used to live in an abandoned one-room schoolhouse way back in the north woods. His mode of travel was a one-horse buggy. He worked by the day at various farms in the area, helping with the haying in summer, or cleaning stables in winter.

He never owned a car; indeed, I don't believe he ever learned to drive one. He got around only with his horse and buggy. But when he retired and moved closer to the village some years ago he sold his horse. Now he walks half a mile to the store to get his groceries.

He was a great walker; even after he got rid of his horse he thought nothing of walking five miles north to help old Bessie Ferguson dig potatoes; or visit one of his former employers. Any season of the year you could see his tall, gaunt frame striding along the country road. Once, in the fall, he stopped in my barn to ask if I could spare a barn cat.

"I need one in the worst way," he said. "The rats are some thing fierce. They walk right across the floor in the kitchen."

It was no wonder. The house he first moved into when he retired was no more than a shanty. He was too poor to have anything else, and of course there was no retirement pay for old retired hired men. He did manage to get a little help from social services.

When the rotting roof of the shanty finally fell in, he moved down the hill to another dwelling and, in so doing, moved up in the world, so to speak. The house he moved to had been vacant for over a year. Its former tenant, Harry Grizzle, another retired bachelor, had suffered a stroke and moved to the county home.

The house was a well-built structure, framed with fading white clapboards. It had a good sound roof, and it was heated with bottled gas. No more did Bill have to chop kindling to feed the wood stove. And, wonder of wonders, it even had storm windows.

And it had a lawn. Bill kept the lawn mowed too. And the better surroundings seemed to improve his appearance. Of course, he never would be listed as one of America's ten best-dressed men, but the hand-me-downs he wore lent his gaunt frame a certain quiet dignity. Even when he smiled, displaying blackened teeth, his face had a rugged nobility. A lifetime of self-reliance showed in the lines etched in his face.

It's too bad Bill was down in the village getting his bag of groceries when the woman stopped to take a picture of his chicken. I'm sure his face would have lit up in a grin to see that.

BARNS INTO HOMES

Seth Cadbury was jubilant. "Do you remember that old barn that stood over there on the other side of the new free-stall barn? The one that we kept heifers in until we got fed up with it being so unhandy and decided to tear it down and put up a pole barn in its place?"

I did remember. It was the original dairy barn on the place, a timber frame structure about thirty feet wide and sixty feet long; they had milked in it right up until the year Seth's father retired. When Seth took the place over and expanded the herd that old barn was too cramped, the stalls too short for the bigger cows, so he built the free-stall barn. For a while they raised calves and heifers in the old barn, but even for that it was unhandy because the drops were too narrow for a gutter cleaner, so they had to shovel manure by hand. Now the old barn was gone.

"So you tore it down?" I said.

"Nope." Seth beamed. "Some guy from Connecticut stopped here last month and bought it for $7000. He brought a couple of other guys; they tore it down board-by-board and hauled it away. Said he was going to turn it into a house. Fancy that!"

When Seth told me the name of the man it rang a bell. I had heard of Edwin Cady. He owned East Coast Barn Builders; he specialized in turning old barns into houses for affluent New York City people who yearned to be weekend country gentlemen.

Now that all the farmers had been squeezed out of Connecticut by high land prices as that state evolved into a bedroom suburb of Manhattan, many of those farmers had emigrated to New York State. The old barns they left behind were reconstructed into lavish weekend retreats for wealthy commuters seeking refuge from city strife. What the rich man wanted was not a conventional second home, but something built on a baronial scale, with hand-hewn beams and the spacious cathedral feeling of high rafters.

Soon all the old barns in Connecticut were used up. The supply shriveled up, but the demand was stronger than ever.

In stepped Mr. Cady, a builder from Long Island. He saw the possibilities and began importing old barns from New York, Pennsylvania, New Jersey and Vermont. New York barns are more desirable because they are larger, sturdier and better suited for conversion to luxury housing, according to Mr. Cady.

About five years ago when he first started this sideline, Mr. Cady could tear down, deliver and reconstruct a barn for from $5000 to $15,000. Now the price is $25,000 and up, just for the bare barn. The option of refurbishing the barn into a house can increase the cost by $100,000 to $1,000,000 or more.

Mr. Cady's business is backlogged for a year.

Most of the barns have been rebuilt in Litchfield and Fairfield Counties in Connecticut. A few have been hauled to the Hamptons on eastern Long Island.

E. Paul Herbert, a publisher, owns a barn in Roxbury, Connecticut that came from the Township of Florida in New York. He says it has a lot of open space, and that's what he likes, not cluttered little rooms..

A barn from Fort Ticonderoga, New York is now in Washington, Connecticut as the weekend home of Norman Pearlstine, managing editor of the Wall Street Journal. He says, "What I was looking for was a fantasy house rather than a conventional one."

But some people are alarmed at the barn traffic. They feel that old barn architecture many become an endangered species and that we are loosing too many of these one-hundred and two-hundred-year-old barns.

"No one knows which barn is historic," says David S. Gillespie, director of the New York Bureau of Historic Preservation Field Services in Albany. He hates to see any of these old barns leave the state. "What makes a building important is the public's ability to look at it in its historical context. If you move a barn to Connecticut, how can the public understand the agricultural life in New York that built that barn?" He makes a good point.

Preservationists also poke fun at the aesthetics of rebuilt barns because, by the time they are converted, they are little more than frames on which new houses are built.

But Mr. Cady disagrees. He admits that, to make a two-and a-half-story barn livable, it has to be insulated. Usually the inside walls are covered with sheetrock and plaster. Sometimes the old beams cannot support the weight of a contemporary house and have to be buttressed with a second frame.

Other than that, the barns are the same. The exterior of the new 'barn-house' is finished with the original barn siding. "We put it back exactly as it was," Mr. Cady says. "Besides, some of these barns were going to be burned anyway."

Mr. Cady keeps several unbuilt barns stacked away in storage as a precaution against the barn shortage spreading to New York and other states. He reckons he has bought and rebuilt about two hundred abandoned or unused barns so far.

He always takes a photograph of an old barn before he tears it down. That way a prospective buyer can pick out his dream house while it is nothing more than a snapshot and a pile of old lumber. Other times he rebuilds a barn and then waits for a buyer.

"I once had five guys standing in the driveway of a barn I had just built, all offering me bids on it," he said. "But another guy had just mailed me a check from France. He got the barn."

SAM the COW-MAT MAN

Mildly curious, I watched as Sam placed his wooden tool box on the barn floor and bent down to unfasten the lid. Curiosity turned to astonishment when the lid flipped open. The box was crammed with knives; knives of every shape, length and description.

Sam tipped the box over, allowing the knives to clatter helter-skelter on the concrete floor. He rummaged among them, pushing aside wooden-handled kitchen knives with ten inch blades, bone-handled steak knives, and smaller knives with narrow blades like stilettos. At the bottom of the heap he found the object of his search. He straightened up, holding in his hand a butcher knife with a fifteen-inch blade. To me it looked as mean as a machete.

"Ah. Yes. This is just the one we need," he said.

He tested the keenness of the blade with his thumb and squinted along the edge with a critical eye.

"This wants to be dressed up just a hair."

When he found his whetstone and began stroking the blade along it, I was reminded of that scene from Alfred Hitchcock's thriller, 'Rear Window', where the killer stands at the kitchen counter sharpening the carving knife.

But Sam wasn't going to carve up my cows. When the butcher knife was sharpened to his satisfaction, he walked over and kneeled down on the black rubber mat and began slicing along the chalk-line. The rubber strip peeled away like butter.

"All right! Okay! He breathed a contented grunt. "That's the way it should cut."

We each grabbed a corner of the heavy mat and dragged it across the barn floor gutter; then we slid it on to the stall platform. A perfect fit. But, of course, you would expect nothing less than perfection from a craftsman. And this was a real craftsman; he was known all along the east coast as 'Sam, the cow-mat man.'

His line of work, spanning a quarter of a century, was installing custom-fitted rubber mattresses for cow stalls, horse stalls and milking parlors.

I soon learned what the narrow-bladed knives were for. The next mat we worked on had to have a semi-circular hole cut into one edge to fit around an inch-and-a-quarter pipe stall divider. The narrow blade was ideal for that work, being flexible working in a confined curve; in Sam's grip the inch thick rubber yielded to the thin sharp blade, leaving a perfect half-moon that fit snugly around the post.

But why did a man need a tool box full of knives by the dozens? Surely two or three would serve the purpose. I asked Sam why.

"Oh, I guess I just collect knives, the way some men collect guns, or housewives collect kitchen trivets. It's gotten to be a habit with me. I'm always looking for the ideal tool for the job. Now, look at this blade; I found it at a garage sale. And that one I bought at a flea market."

To Sam, a cutting edge had an infinite allure. I guess it's a part of our human culture; a trait probably originating with the cave man who first chipped a piece of flint to a sharp point, and later learned to grind an edge on a sharpening stone. Knives have been an essential tool of mankind since the Stone Age.

Casting another glance at Sam's assortment of knives, I could better appreciate his affection for them. There is nothing more important than the proper tool for the job, and nothing more frustrating than an improper or dull tool.

Roughly half of his knives had seen earlier days in a kitchen. There were knives for slicing, carving, paring, peeling and chopping. Other had seen service in the carpenter's trade, and some in the shoemaker's line of work. There was one that looked like an oilcloth knife, and another that probably had been used to open oysters. I'm sure that Sam had a need at one time or another for each one of them — that special cutting job that needed a special shape of blade.

I imagined that collecting knives would be hobby enough for one man. But as we worked, fitting the rubber mats, drilling holes through them into the concrete platform and fastening them with nylon anchors, I learned that Sam's real love was gardening. He looked forward to spring when he could get down on hands and knees in his garden plot and probe the good warm earth with a trowel, planting seeds.

"There is absolutely nothing more relaxing than working with growing plants," he said. "After a couple hours' work in the garden, the cares of the day just melt away. It's better than any tranquilizer could be."

He spoke with reverence of the tremendous vitality of Mother Nature. The generative qualities of land, under the benevolent warm sun and warm showers, inspired him. Waiting for seeds to sprout, and watching the sturdy seedlings shouldering their way through the earth crust was a joy whose luster never faded.

He said that as the seedlings grew taller, and the sap pulsed through elongated stems and leaves, he felt their life force almost as a real vibration. Touching the leaf of a corn plant for instance was like grasping the hand of a friend, and feeling a clasp in return.

"After manhandling one-hundred pound cow mats all day, and carving tough rubber with a knife, it's relaxing to work with tender plants and yielding soil," he said.

I could understand that. I also could feet rapport with a man who loved the earth so strongly that he wanted to absorb nature to the last precious drop.

THE CATTLE DEALER

I promised myself that if I ever got out of the war alive, what I wanted to do when I got back home was get a little cattle truck and buy and sell calves and cows for a living," said Ray.

Ray Nagele made it back home alive in October 1945, although the beri-beri he fell ill with at the Japanese prison camp almost claimed his life. The deficiency disease made his body grossly swollen so that he looked like he weighed 300 pounds, although he actually was at a starvation weight of 100 pounds.

When his health returned he did become a cattle dealer; he enjoyed that line of work for nearly forty years. He stopped at my farm one day to show me the document he had just received from the United States Navy. It was an honorable discharge, issued thirty six years late; Quite a tale is connected with it.

Back in 1941, Ray was a nineteen-year-old boy fresh off the farm. A buddy of his was working on construction for the Morrison-Knudsen Contractors; the project was repairing one of the canal locks on the Mohawk River near St. Johnsville.

The Morrison-Knudsen firm had another work agreement with contractors for Pacific Naval Air Bases. Ray had a chance to go to Wake Island along with 1,200 other civilian construction workers employed by Pan American Airways. The job offer was as a driller's helper.

"Me, a green farm kid," Ray recalled, smiling, "When they said 'driller's helper' all I could think of was drilling oats at planting time.

Ray and a couple of his buddies went to Wake Island and worked for six months. At the end of the summer they had the option of signing on for another six months, or going back home.

"Why do we want to go back home in the fall?" one of his buddies asked. "We'll just have a cold winter to face. Let's work here another six months and go back home in the spring."

So they stayed. They were there when war broke out in the Pacific following the December 7 bombing of Pearl Harbor. Then it was too late to go home. The construction workers were civilians, but many of them volunteered to help the United States Marines and the Navy defend Wake Island when the Japanese attacked it shortly after Pearl Harbor.

There were about four hundred Marines and one hundred Navy men on the island at the time. They asked the construction workers for help in defending the island, and Ray was one of many who volunteered.

"The garrison finally surrendered to the Japanese on December 23, 1941 and I was taken prisoner along with the others," said Ray. "I remained a prisoner until October 10, 1945. I was in five different prisoner-of-war camps: Woo Sung and Kai Wang in China; Osaka, Kobe and Naoetsu in Japan."

Cruel treatment and beatings marked most of his days as a prisoner of war.

"When they took us prisoners by ship from Wake Island to China, it was January and very cold," Ray recalled. "We had nothing to keep us warm. On the deck there was a pile of about a hundred blankets, but we were strictly forbidden to use them. The men would sneak the blankets at night and return them before dawn.

"One morning the Japanese caught us. We had to line up and each man got five whacks on the head. I was so tall they barely could reach my head, so I got two whacks on the head and two on the back."

In prison camp in China, the 1,600 prisoners had to grind flour by hand. The main topic of conversation in prison camps for the entire four years was food. Most of the time their diet consisted of a rice ball and some seaweed.

"When we were released at the end of the war and got back to San Francisco, the first thing we did was go to a restaurant," he recalled. "My buddy and I had a terrible craving for butter.

"The waitress said there wasn't any; everything was rationed. We told her we had just been released from four years in a Jap prison camp. She said 'Wait a minute' and came back with a full pound of butter on a plate. Of course, we could only eat a little dab of it, our stomachs were so shriveled up from starvation all those years."

From China they were transferred to Osaka camp in Japan in August 1943 to work in the shipyards. They had to work in the rain, wearing raincoats.

"I caught red-hot rivets in a funnel when working at the yards and used tongs to hand them to the guy working the rivet gun," he said.

Ray can still count to one hundred in Japanese. And he remembers certain words: *yasume*, which means rest; and *ohio*, which means good morning.

A wry smile flits across Ray's face with the remembrance of scenes past.

"I can still hear the Japanese as they came to work at the shipyard each morning. They would politely say good morning to each other: *ohio, ohio, ohio*—it was a regular chorus all along the way."

Ray recalled that the prisoners had to walk five miles from camp each day to work in the shipyard, and five miles back.

"We marched four abreast," he said. "It was part of the Japanese training."

In January 1945 Ray's beri-beri got so bad that he was trucked to the Kobe hospital in early February.

"It took all night to go thirty miles," he remembered. "There I was, standing on the back of the truck, and a guard had his bayonet pointed at me all night long."

While in Kobe he came down with appendicitis; his appendix was removed on May 1, 1945. "The operation was performed by an English doctor who had been captured in Singapore," he said.

"Five days later I was sent by train to northern Japan, across from Siberia, and stayed there until the war ended. I was put on light duty, sweeping in a steel mill. I got two beatings there," he said.

The war in the Pacific finally ended August 15, 1945. Nagele was released in October 1945 and brought back to California. His mother and sister flew to the west coast to meet him and accompany him home in his debilitated condition.

At Honolulu in 1975, thirty years after his release from prison camp, there was a reunion of the survivors of the Wake Island 'Naval Civilians'.

Six more years elapsed before their petition was acknowledged by the U.S. Navy — that although never formally inducted, they had volunteered to serve with the Marines in the defense of Wake Island, and therefore were entitled to recognition and an honorable discharge.

"Better late than never," Ray said with a slow smile.

Then his smile faded and he said, "War is a terrible, awful thing. There's nothing more precious in this world than freedom."

Freedom to Ray Nagele is the little red cattle truck he dreamed about during almost four years in a Japanese prisoner-of-war-camp.

THE OLD JEEP TRUCK

It was November when I faced the question of whether my farm pickup truck would make it through the winter. The truck was a 1957 Willys Jeep that I bought from my oldest brother when he retired and began spending his winters in Florida. It was a rugged workhorse for plowing snow. With our Mohawk Valley winters a farm needs a tough vehicle for clearing the deep snowfalls that bury our farm driveways.

The speedometer had been completely around once and registered more than 20,000 miles on its second circuit when I took possession. Although the front half of the body was in good condition, the rear fenders were nearly rusted off as were the bottom panels under the doors.

It lacked the sleek round lines of the brand new trucks everyone else seemed to be driving. Indeed, it looked like a box on wheels. And it never failed to elicit comment when I drove in a neighbor's yard.

"How old is that truck anyway?" would be the inevitable question. My standard reply was, "It's been old enough to vote twice over."

That November day I stood back and took a long hard look at my truck. I scrutinized the rusted spot on the frame that I had reinforced by welding on a used plowshare. The bottom half of each door was heavily speckled with rust. Like every vehicle exposed to northern winters, it suffered rust damage from the heavy use of rock salt on our winter roads.

The doors had a habit of popping open whenever the truck went over a bump, but I kept them tied with baling twine. The truck really needed new shock absorbers.

As I stood there with my chin in my hands it seemed that the truck with its two headlamps and old-fashioned elongated grill was smiling at me. Should I trade or shouldn't I?

I thought of the money that had to be spent to renovate my barn gutter cleaner and manure storage area. Then I pondered the five-figure price new four-wheel drive trucks were selling for. The decision was easy — the old truck would stay another year.

With a new battery installed the Jeep ran like a top all through December. We had our first heavy snow Christmas week. The Jeep plowed with a will.

Then one cold morning in January as I drove in the yard at

the upper farm to feed the heifers, it happened. I put in the clutch to stop; something went 'snap'; the clutch sank to the floor and the truck stalled.

I got out and crawled underneath to see what was the matter. One of the linkage arms from the clutch pedal had broken, an L-shaped rod about five inches long. Well, that was a minor thing. Or so I thought.

The Jeep garage in the next village didn't have that particular part in stock. I took the broken rod to Harold Burkdorf's repair shop.

"Do you think you can weld this?" I asked.

Harold turned the two broken pieces over in his hand and looked doubtful.

"That's hardened steel and would probably break again. Why don't you try Leland Smith. He just might have that part."

"Who's he?"

"Leland Smith over on Cottage Street. He's a retired mechanic. Does some repair work in his garage. Last year he bought up a whole inventory of Jeep parts when a dealer went out of business. He lives on the corner, a gray shingled house."

I drove over to the gray shingled house on Cottage Street but no one was at home. There was no sense waiting around. I went back to the farm for lunch and later telephoned to see if Leland had returned home.

I was in luck. Leland himself answered the phone. I told him I needed a clutch linkage rod for my Jeep.

"Short wheel-base Jeep?" he asked.

"A pickup truck, 1957."

"Oh, a Willys. What engine."

"Hurricane six."

"Hmm. Hmm. Let me see. You want a rod about four or five inches long, bent on both ends. Yes, yes, I got one. I got one."

"Good! I'll be right down."

When I got there ten minutes later Leland's wife opened the kitchen door, and Leland, a thin man bubbling with energy that belied his sixty some years, beckoned me from the back pantry adjoining his one-stall garage. The pantry had been converted to a storeroom for auto parts. Bins lined the walls. Three engines huddled on the floor.

Leland searched in a bin, pulled out a small brown envelope and took from it a bent metal rod. "This what you want?"

"That's it! How much do I owe you?"

"Let's see. Three dollars. And that's August 1978 price, not January 1979."

I handed him the three dollars gratefully.

"Anytime you need parts, I got them." His shiny round cheeks glistened as he smiled. He gestured proudly at the parts bulging from the wall bins. "Transmission parts too. I can do a complete transmission overhaul if you need one."

"Thanks, Leland. I know where to come if I need any more parts."

A week later I was on the phone to him again.

"Leland, I think my clutch is going. The throw-out bearing is starting to whine."

"Chatters does it, when you put the clutch in? Probably one of those fingers on the clutch plate is broke off. I got the disk, and I can get you a plate."

"How much will it cost to put a clutch in?"

"A rebuilt one? Ninety-five to a hundred dollars, or thereabouts. That's for parts. Labor is extra."

"How long will it take?"

"Two days. Two good warm days. I'll have to let the truck set overnight in my garage to thaw out."

That Saturday I was back on the phone again.

"Leland, my clutch is gone."

"I got the parts. I got all the parts. Can you get her in gear and drive her down here?"

"I'll get her there!"

I had to start the Jeep up in low gear and drive it all the way down to the village in low. On Cottage Street I made the turn in to Leland's driveway and shut off the ignition to make the truck stall to a stop. Leland opened the garage door and I eased the truck inside over the pit by flicking the ignition switch on and off.

"You've got no clutch at all?" he said as I got out of the driver's seat. "If it's gone that far it's possible you've chipped a tooth off the sliding low gear. It's possible."

Three days later the job was done. It took an extra day because Leland is a member of the volunteer ambulance corps.

"We had to make two runs with the ambulance, so that took me away for better than half a day," he said. "But she's all done now. You've got a beautiful clutch now—a beautiful clutch."

He led the way to his workbench. "Here's the casualties, your old parts. I want you to see what they look like:

"Your old brake disk— gone.

"Clutch press plate— both fingers broke off.

"Throw-out bearing— really worn.

"Low and reverse sliding gear— see how the teeth are worn?

"And here's your itemized list of parts: Seals. Spring. Carrier and sleeve. Eleven pounds of gear oil for the transmission and transfer case."

"Did you have much trouble getting it apart and back together?" I asked.

"No. It's no problem when you know how to work these babies. The secret is to have a dummy shaft to use when you pull the transmission. Then you have absolutely no problem. No problem at all. See — I've got a shelf full of dummy shafts for different sizes."

I wish I had had a tape recorder as he went on explaining his procedure. The gems of knowledge from old-time mechanics such as him, with all his tricks of the trade, should not be lost to posterity.

After I wrote out the check and got ready to leave, Leland looked at the truck affectionately.

"That Hurricane engine is a good one. You can't hurt them. I'd like to have one with overdrive. That's the old Kaiser engine. Yessir—the old Kaiser engine."

For a moment it looked like he was going to pat the hood.

"You've got a beautiful clutch there. But if you notice the least bit of tightness in the clutch pedal, bring the truck right back and I'll adjust it."

I backed out of the driveway and headed up the street toward home. It was a beautiful clutch. Smooth and firm. And it was an indestructible truck. Why, with new shock absorbers and two new rear fenders, it could go another hundred thousand miles!

KILLDEER

Now in November, we're in the twilight of the year. These mornings at six o'clock when I walk outside toward the barn for the early milking it's still dark, with only a pale yellowish tinge in the east to show where the sun will come up.

The countryside is hushed this early in the morning, this late in the year. No pre-dawn serenade of the robins that set the June mornings throbbing; the robins have all gone south, as have all the other songbirds, weeks ago.

But for many mornings now at six o'clock I've heard one summer bird. High up in the dark sky a killdeer calls. Actually there's a pair of them. I can't see them but I can hear their plaintive cries — killdee, killdee — as they glide somewhere under the stars, high above the barn roof.

Why haven't they gone south, as all the other birds have done? Why do they linger behind? I wonder if they are the same pair of killdeer that nested in the cornfield? Are they reluctant to leave behind all those fond memories?

I always have to pause halfway to the barn and listen to them. They keep circling up high and keep calling, for more than an hour, until the sun comes up. Then they land somewhere, to feed no doubt, and are quiet the remainder of the day. But always they are up in the dark sky again the next morning.

As their calls drift down from the dark sky, I seem to read in them all that has transpired during the past summer. They tell me that the leaves are old and fallen, as if I didn't know. And they say, remember the early petal fall in June when pear and apple blossoms went down in showers?

And I do remember. The blossoms looked like snow on the brown earth of the cornfield; and the corn itself was just coming up, the rows of tiny plants looking like strands of emeralds on the dark ground. The killdeer hen bird was nesting near the front end of the field, warming four bluish eggs splotched with brown; the killdeer eggs were so heavily speckled they looked indistinguishable from the pebbled nest on which they lay.

Remember the flowers? Or so the killdeer seemed to say, that for flowers mid-June is to November as one-hundred is to one. There aren't any flowers in November, unless you count the witch-hazel bush, the only shrub that blooms in autumn after all the leaves have fallen.

But June! Columbines and daisies! Solomon's plume and

miter wort! Foam flower and roses, bell flower and lobelia! Every day in June at least four new flowers opened their petals for the first time. The day I noticed the first blue lobelia was the day the killdeer eggs hatched. I chuckled with pleasure when I saw the hen bird trotting among the foot-high corn plants, followed by four precocious fuzzy chicks like mice on stilts.

And when a sudden shower sent raindrops pelting down, spattering the dust, the chicks rushed for cover under the outstretched wings of the hen bird. And at night there were those same wings again, that same protective warmth.

Yes, the warmth. Remember the warm, benevolent sun the killdeer seemed to say — the sun that made the corn grow so fast in July, You could actually hear the corn grow on hot July nights —that crackling sound was the corn growing another inch.

Ah, those were the days when the warmth brought out the bugs; and the killdeer trotted hither and yon in the cornfield eating up those bugs. The ground under the rustling, shoulder high corn stalks became one vast platter where the killdeer dined a la carte.

Remember the breezes, the winds, the gales? The killdeer remember them all. They remember the warm updrafts that rose high above the cornfield, those warm spires of air on which a bird could float and circle for hours. They recall the day in early August when the four young birds first tried their wings.

For weeks, the young birds had been taking short hopping flights along the ground, much as a person who is learning to fly an ultra-light does. Then one day they took to the air, climbing fifty feet, one-hundred feet above the waving corn, accompanied by screams of encouragement and delight from the parent birds. Higher and higher they soared until they were mere specks in the sky.

Thereafter they spent hours at a time each day in flight. Oh, the joy of it! To spread your wings and climb, and soar, and glide! The six birds called to each other as they performed flight maneuvers, and you could read sheer pleasure in each call.

Now in November the killdeer are calling again, but in a dark pre-dawn sky. But why only two? Ah, there...another call and another, and then two more.

How many days before they will turn and set their course for the south? We've had several hard frosts already, although the days do warm up and the afternoons continue to be sunny, calm and mild. Yet autumn is winding down; nature's clock is winding down. There are fewer hours of sunshine every day, diminished warmth in the paltry sunshine, less of everything. Still, they are reluctant to leave until the very last minute.

They circle and call, and the question that they seem to ask is what to make of these diminished things.

THREE-DOLLAR BILL

The United States Treasury used to print two-dollar bills, but I never saw a three-dollar bill until I met a member of the equine family by the name of Bill. Bill was a brown pony that my farming neighbor, Bob, bought one summer for his kids to ride.

Most people think that ponies, being of small stature, are docile with small children. However, children have neither the physical strength nor the firmness to make a pony obey. When a pony learns that he can get the upper hand, he soon develops bad habits.

Like many ponies that go from horse dealer to family, and back again to the horse dealer, Bill had only received the barest rudiments of training. He would let the children sit on his bare back occasionally, but he didn't like the saddle. He would walk if he wanted to, but if he didn't want to, nothing could make him move.

"Talk about stubborn!" said Bob. "That pony must have been part mule. We used to tie him out on the lawn with a halter and a long rope so he could eat grass. When he got the grass chewed down, I tried to move him to another spot. I pulled and pulled on the rope but he wouldn't budge until he felt like it.

He liked to chase cows too," volunteered young Bobby.

"I put him in the pasture with the cows just to see how he would act," explained his father. "That was a mistake! He liked the freedom of the wide open pasture so he started to run. That made the cows nervous and they began to run. Bill thought it was a game, and he started chasing the whole dairy herd round and round."

"Oh, did Bob get mad!" Pat, his wife, said. "He almost got the gun out to shoot the horse. We finally got Bill out of the pasture and we never allowed him back in there again. He couldn't be trusted with the cows."

The first winter they kept Bill in the barn in a stall next to a white cow. Whenever the cow tried to reach over and steal a mouthful of Bill's hay, Bill bit her.

"He always had a piece of white hair in his mouth," said Pat. "That poor cow!"

During mild winter days they turned Bill loose to wander around in the farm yard and get some exercise along with fresh air and sunshine. Pat started feeding him sugar cubes as a treat.

"That was a mistake! He used to wait outside the kitchen door for me, expecting sugar every time I came out of the house. He used to lie down by the step just like a dog. And if I didn't give him any sugar, boy would he get mad! He would follow me to the barn, trying to nip me, and even tried to chase me."

With his fuzzy winter coat, Bill looked like a bear. In the springtime when he shed that long hair and slicked out, he looked beautiful.

"He was really a nice looking pony," Pat said. "It's a pity we didn't have the know-how to train him properly."

It was when Bill started to chase the kids that they decided to get rid of him.

Admittedly, late fall is a bad time to sell a horse. Everybody wants a horse in spring and summer when the weather is fine for riding. Nobody wants to feed one all through the winter.

Still, when Fred, the horse dealer, came they were dismayed at his offer. Fred zipped his jacket up against the cold west wind and eyed the pony morosely for several minutes. Finally he spoke.

"I'll give you three dollars for him."

Bob exploded with indignation.

"Three dollars? I'll shoot him before I sell him to you for that!"

Fred turned his sad eyes from the pony to Bob. "This time of year most people are giving ponies to me just so as I'll take 'em off their hands."

"Well this is one pony I'm not going to give to you. Three dollars! That's an insult!" Bob stalked away. The horse dealer left.

It was outrageous, really. Not that they expected to make any money on the horse. "For heaven's sake! He's worth more than three dollars just for dog food," said Pat.

Bob eventually talked to Murdock, the cattle dealer. He agreed to take the pony and try to sell him as best he could, and they would split the money. Murdock finally found someone willing to buy the pony for seventeen dollars.

Bob and everyone always remember that pony as "Three Dollar Bill".

CORN

Three ears of multi-colored Indian corn, braided by the husks and hung by the front door, signify the end of the harvest season. For the homeowners of suburbia, who neither sow nor reap, the corn nevertheless serves as a reminder of an agricultural heritage. For me as a dairy farmer, corn is an indispensable feed for my cows. Corn is the most important grain grown in the world today, second only to rice. We owe it all to the American Indian plant breeders who developed it centuries ago, probably from the grass known as teosinte.

Multi-colored corn was just one of at least six kinds of

corn planted by the Pueblo tribes of the southwest. The first time I saw blue corn was in northern New Mexico. A little restaurant there served blue corn enchiladas, with red chili sauce. Compared with the enchiladas they served back east, it was like the difference between vine-ripened tomatoes and the tasteless kind supermarkets sell in winter.

Also on the menu were blue corn burritos with black beans and rice, topped with guacamole, sour cream and green chili sauce. Needless to say, I sampled those, as well as blue corn tacos stuffed with chicken and avocados.

Having eaten these Indian breads made with blue corn flour I was curious to see what blue corn looked like on the cob. On the rooftops of the pueblos there were newly-harvested ears drying. I saw that the color of the kernels sometimes was a true blue. More often it was a purplish-black, the color of winter storm clouds, or an iron gray.

For centuries the tribes have cultivated blue corn on the high, dry mesas and river valleys of northern Arizona and New Mexico. When the conquistadors arrived in the New World, they adopted blue corn flour to make that staple of the Spanish diet — tortillas. In turn, tortillas were adopted by the Indians to supplement their various boiled breads and dumplings.

Like all breads, blue corn tortillas taste best when eaten the day they are baked. Since few restaurants make their own tortillas, the best blue corn specialties are found not far from a blue corn tortilleria — a place where tortillas are made. That's why blue corn baked goods rarely are found outside the southwest.

Like all corns, blue corn is a maize—from the Spanish *maiz*. The name probably came from the Mayas, the ancient Indian civilization of Central America. Several primitive varieties of corn have been recovered from the most ancient tombs of the Mayas. Hence the botanical name, Zea mays.

Blue corn is a 'flour corn'. The kernel is hard on the outside but soft and white on the inside, and starchy rather than sweet. After the ears are dried on the rooftops, they are shelled as needed, and the kernels ground by hand into the gritty, lavender corn meal or flour that gives blue corn tortillas their characteristic color and texture.

Blue corn is even used to make waffles, muffins, pancakes and chips. I also tasted a blue corn pizza dough. Most often though, blue corn appears as a tortilla. For a great way to start the day I had breakfast of huevos rancheros — eggs on top of tortillas, with chili sauce.

The Pueblo tribes grow blue corn, as well as white, red, yellow, black and multi-colored. Traditionally, blue corn is planted by hand and cultivated by hand. Since it has low standability — the weak stalks tend to topple over with the weight of the ripe ears — it is also harvested by hand. Consequently, it is scarcer and more expensive than other types of corn.

The Pueblo Indians use it to make a hot cereal called atole. They also use it to make a very thin, multi-layered and rolled bread called 'paper bread'.

The Indians grind blue corn meal on a large stone slab called a 'metate', using a smaller hand-held stone called a 'mano'. In days gone by, the skill with which Indian maidens ground corn meal was used to judge their industriousness and hence, their marriageability.

Of all the breads the Indians make with blue corn meal, paper bread, or 'piki' as the Hopi call it ,is the most traditional and the least adaptable to modern life. Thus it remains a link to the ancient past. It is cooked over an open fire on a large flat, glossy-black stone that is handed down from mother to daughter. The hot stone is greased by smearing it with sheep brains; then a thin layer of batter — corn meal, ash and water — is spread across the stone by hand. The batter cooks quickly into a paper-like sheet which is peeled off the stone, folded and rolled.

On the Hopi reservations of northern Arizona, piki is still made to celebrate weddings and ritual dances. When you see smoke rising from the chimneys of piki houses, you know a ceremony of some kind is only a day or two away.

Blue corn meal is used in a 'naming ceremony' for a new baby, in death rituals, and in the dances that accompany the seasonal prayers for rain. One anthropologist, Richard Ford, believes the various colors of corn helped the Pueblo Indians survive the wayward weather of the high plateaus of the southwest.

Ford reasons that, in order to preserve the purity of color of the different strains of corn, the fields had to be planted far apart to avoid cross-pollination. The unexpected side effect of this was the lessening of risk for total crop failure. If one field was lost by hail, or frost, or flood, at least some of the others were likely to survive. Thus the Indians always had food to get them through the winter.

For the Indians, corn has a religious significance. Pueblo people believe their ancestors emerged from the spirit world with corn seed.

Whatever its ancestry may have been, the cultivation of corn by the Pilgrims helped the struggling colonies to survive.

The stony forested land of New England was poorly suited to the cereal crops the settlers had been familiar with in Europe. The Indians not only gave them ears of corn but showed them how to plant it.

Corn was easier to grow than the Old World grains; it had a more abundant yield and was ready to eat at an earlier date. In addition, it furnished more and better fodder for the colonist's animals; the stalks and leaves could be fed to livestock. The husks could be used for mattress stuffing.

On their first Thanksgiving Day, in December 1621, the colonists of Plymouth who had survived the first terrible winter gave thanks for their harvest of corn.

FARM ART

In a city within forty miles of my farm there is a fine museum of art that my wife and I visit once a year. We went recently to view their latest exhibition: a collection of paintings and sculpture by artists residing in upstate New York.

The art museum is an impressive new building designed by a renowned architect. Their permanent collection of art work is equally impressive. They boast, among other things, several paintings by Winslow Homer, a Renoir, a Degas and a Picasso.

The current exhibit by contemporary regional artists was not in the same league as those masters, but it was interesting.

There was the usual motley group of abstract paintings in vivid acrylics that constitutes one branch of modern art; also an innovative and sometimes weird assortment of photograph collages. But the gallery with sculpture was something else.

I had to laugh to myself — a quiet, museum laugh — when I viewed the first piece of sculpture. A small white card gave the name as 'Untitled.' It consisted of pieces of hardware cloth rolled up to form sixteen cylinders, each about four inches in diameter and four feet long. These wire mesh tubes were laid flat, placed four in a row and four tiers high to make a compact bundle; the complete bundle was held together by a couple of strands of smooth wire wrapped around the whole thing.

With a few minor adjustments that 'sculpture' would be as nice as a similar contraption we have at our farm — a sparrow trap. We use it for catching those pesky English sparrows that chase bluebirds away. Our sparrow trap is a wire mesh box made out of hardware cloth identical to that of the sculpture; the sparrow pathway into the trap is a cleverly made tapered tube of wire mesh; the tube is wide at the outside entrance but narrows to a small opening inside. We bait the trap with cracked corn. The sparrows find their way in through the tube but aren't smart enough to figure their way out.

After I had looked at 'Untitled' from all angles, I came to the conclusion that our sparrow trap was more artistic. I moved on to the next sculpture.

This one was also named 'Untitled.' It was composed of four weather-beaten wood fence posts holding up a two-foot square section of old fashioned woven wire guard railing (similar to cyclone fencing). The guard rail wire had come direct from the highway roadside where it had obviously fallen and lain for some time: it was thoroughly rusted and had a thick coat of dust and a few clumps of embedded sod.

I had no trouble at all thinking up a suitable name for this one: 'Hog Pen.'

I could hardly wait to get to the next pedestal. I knew I wouldn't be disappointed, and I wasn't. This third piece of sculpture was also 'Untitled.' That seemed to be the fashion in names. It was a brand-new, black metal, six-inch-diameter, stove pipe 'T'. That is correct — a stovepipe 'T.'

I can't for the life of me understand why the 'sculptor' couldn't think up a title for that one.

In high spirits I moved on to Exhibit Number Four. Its name? 'Untitled' of course.

Ah, but this one was different. This one was 'Untitled Number Two'. It was a flat section of boiler plate — a piece of metal two feet wide by four feet long — on top of which rested a rusty metal wheel from some old farm implement.

Really buoyed up, I continued to Exhibit Number Five. This was a wooden lobster trap, apparently salvaged from a junk heap. Most of the wooden mesh sides were broken and missing and had been replaced with woven baler twine. It reminded me of a gift my nephew made for his mother as a sixth grade project: A tin can with yarn wound around the outside; it was meant to be used for displaying Christmas cards.

So this was the state of the arts. Inspired by what I had seen at the museum, my mind began searching for what I could contribute. This occupied me on the drive home.

If junk and scrap metal could be made into works of art, my farm was a potential gold mine. The scrap metal heap I have in the old shed — I'm sure every farm has such a junk pile — is full of odd bits of angle iron, links from the barn cleaner chain, broken springs, odd sprockets and what-have-you. There may be some odd thing in a scrap pile you never will use in twenty years, but the day after you throw it away you find you are in desperate need of that exact thing.

Any two pieces from that junk heap, fastened to a block of shiny walnut, would make a fine sculpture. The more random the selection the better. I could just close my eyes and pick up the first two hunks of metal I laid hands on.

The next few days, as I was doing my chores around the barn I kept my eyes open for possibilities. I really began seeing things for the first time. Or to put it another way, I began to see things in a new light.

Here was that old metal swing stanchion that broke at the top and that I heaved in a corner 'temporarily', and here also a rusted-off and bent stall-divider pipe. Artfully positioned together, what a perfect 'Untitled' they would make!

And this ancient section of spring-tooth harrow laying by the old stonewall — there is another potential masterpiece. Not to mention the tongue off the old forage harvester, bent in such a lovely shape when the wagon jackknifed—that is for certain a 'Best of Show'.

But it was not until I walked in the heifer yard that I found my chef d'oerve. There in the salt box, lovingly sculpted lo these many months by loving tongues — the fifty-pound block of salt! It was molded in undulating curves and ultra-smooth depressions. It was a thing of beauty. Not one line dared be changed. I grabbed it before the heifers could ruin it. You know how some artists don't know when to stop. Just wait until the panel of jurors at the museum sees it.

RUSSIAN TRUCK-JUMPING

Over the years, I have towed many a vehicle out of the snow and mud with my farm tractor. An instance that comes immediately to mind happened on a hot June afternoon when we were baling hay in our furthest field, behind a woodlot. Although we were out of sight and out of sound of the barn and house, the man in distress found us.

Rusty was on top of the hay wagon and spotted him first— —a weary figure trudging along in hip boots. He was a fisherman who had pulled his car off the highway to park near the brook. Too late he discovered that the tall weeds at the edge of the road hid an open ditch coming from the road culvert.

In gratitude for pulling his car out of the ditch with our tractor, he gave us fresh trout from his creel for our supper.

And on a certain December day, even our mailman veered off the road into a deep snow-filled ditch. He had been driving with one hand and sorting mail with the other — not a smart thing to do when the winter roads are slippery. The U.S. Mail was buried so deep, it took my biggest four-wheel-drive tractor with chains on the rear wheels to pull him out.

The next morning in my mailbox was a bottle of Christmas wine.

But this time the shoe was on the other foot. I was stuck in the mud, or to put it more correctly, we were stuck in the mud. Stuck, but good; buried to the rear axle.

The twelve of us, three Americans and nine Russians, got out of our vehicle — a two-and-a-half-ton, four-wheel-drive Russian army truck — and took stock of the situation.

We were in the wilds of Siberia, miles from nowhere, homeward bound from a two-week wilderness backpacking trip. It wasn't through any fault of our youthful driver, Slava, that the truck got mired in the mudhole. Although he was only twenty-two, Slava was a good driver — he acquired his skills in the former Russian army — and on the outward-bound portion of our trip the truck never once got stuck, although the roads were horendous.

Our journey on the first stretch of blacktop road, from Irkutsk to Kultuk had been merely bumpy. But the unpaved road through the Tuniskaya Valley to the hamlet of Orlik was your washboard, jar-the-fillings-out-of-your-teeth type of road. For Slava and two other members of our party riding up front in the cab, the rough ride was bearable; but the nine of us huddled under the canvas roof in the rear of the truck, crammed in amongst the backpacks and duffel bags — we felt every jolt magnified. It got so we could anticipate the upcoming mudholes because Slava always gunned the motor for enough speed to plow through. We nicknamed it 'Russian Truck-Jumping.' And from Orlik up into the mountains, the unpaved road swiftly dwindled into two ruts.

Through bogs, across gravel-bottomed rivers, and up into the forest on a logging trail over boulders and fallen trees, the truck lurched and bounced, but never got stuck.

Despite all those hurdles, we made it to our destination — the headwaters of the Sentza River near the Choygan-Daban Pass and the hot springs sacred to the Buryat and Tuva native people. It was truly a pristine wilderness area.

Using a log cabin at the hot springs as a base camp, we spent unforgettable hours hiking with day packs into the wildly beautiful birch and larch forested mountains.

Now, on the last leg of our return journey, after surviving all previous hazards, our truck had failed us by burying itself in the mud.

We weary campers stood in a circle around the truck, surveying our predicament. We were still in a remote area, more than one-hundred kilometers from the nearest village. It might be days before another truck happened by.

Our only hope was the possibility of finding an isolated Buryat herdsman who might have a team of horses or a tractor. But in which direction should we go searching for help?

To either side of the rutted track were occasional trails that the nomad herds followed. We knew that a few of the herders had built log homesteads in the random meadows along the river. Perhaps we could locate one of them.

Fortunately, one of our Russian guides was a marathon runner. Alexander Kitov — his nickname was Sasha — was born in Irkutsk near Lake Baikal; he said he had begun running in the ninth year of school. He had run on the ice in the Lake Baikal marathon. His latest accomplishment had been a fifty mile marathon around Moscow..

So our other guide, Alex Galitsky, handed him one of the walkie-talkies, and Sasha started off at a lope. We settled down to make a campfire and tea.

An hour later, Galitsky's radio crackled; Sasha's voice came over with good news. He had found a farmer with a tractor and was on the way back to us.

In due time we heard the tractor approaching, and soon a blue Belarus drove into sight, coming along the trail through the tamaracks. A chain was fastened to our bumper, and out of the deep hole, mud spinning from its tires, came our truck.

The Buryat farmer was in his first year of privatization since the demise of the collective farms. He was very glad of being his own man again and also pleased at having pulled these crazy Amerikanskys out of the mud.

I knew only a few words of Russian which I pronounced to tell him I was a farmer too.

He knew but four words of English, and they were eloquent: "I Buryat," he said, and pointing to the Belarus he added, proudly, "My tractor!"

DYNAMITE

Frank Windsor was going to put up a fence that would last a long time. The fence between the cow pasture and the farm house lawn presently was a woven wire stock fence, but it was sagging and stretched out of shape from the cows reaching under it; and the whole line of fence posts was leaning inward.

The problem was in the ground itself; there was solid rock less than two feet below the surface. You can't set fence posts only two feet deep and expect them to remain solid, not when thousand-pound cows strain against the wire.

In our part of the country you have to dig a hole for posts

that will reach below frost level. If posts aren't driven deep enough, the thawing ground in springtime will heave posts up several inches. After several years of such freezing and thawing, the posts become loose; all a cow has to do is lean against such a post and the fence will topple down. You must have a tight fence for cattle.

Frank knew this, of course. But what do you do when ledge rock underlays much of your ground? It was frustrating trying to drive posts in such a situation. He had suffered with that weak fence long enough; the cows had gotten through it and on to his lawn one too many times.

Frank was a perfectionist; a crooked fence was an abomination to him. He wanted the fence around his farmhouse lawn to be straight, sturdy and a thing of beauty, and he wanted it to last a long time; so Frank decided to put up a post-and-rail fence. The posts would be locust, a hardwood impervious to rot; they would last practically forever. And to make certain the posts were anchored deep enough, Frank and his hired man were going to blast holes for each one.

Bertram, the hired man, was a jovial middle-aged fellow. He had a perpetual amused look on his face and could find humor in any situation. He also was shrewd and practical. When Frank announced that he was going to use a stick of dynamite to open up each post hole, Bert pursed his lips and squinted his eyes in thought.

"What about the windows in the house?" he asked. "A blast of explosive is going to send pieces of rock sky-high. One of them pieces is bound to crack a window, sure as shootin'."

The house was about a hundred feet from the fence line.

"Don't worry, I've already thought of that," Frank said.

"We'll use that old tailgate from the truck to cover each hole. That will keep the rock from flying too high."

Bert thought this over and grunted his approval. "Um, okay. That oughta do it."

So they went at it. They strung a line out and every ten feet they dug down with a shovel until they hit rock. Then they drilled a small hole in the rock, just deep enough to hold a small stick of dynamite. They packed mud around the charge, and set the old tailgate over that, and moved back to a safe range while they set off the first charge.

Whoom! The ground shuddered. The metal tailgate rose twenty feet up in the air, pushed up by a fountain of dirt and shattered rock. Then with a dull thud the tailgate dropped to the ground, somewhat more dented than it had been at the start. It did the job perfectly; no flying rock got anywhere near the house.

It was slow, laborious work, digging down through the dirt, then drilling into the rock for each charge of dynamite. But they stuck to the job, and by lunch time they had a dozen holes blasted and a dozen locust posts set, each one four feet deep.

After lunch they went right back at it, for there were fifteen more holes to blast. By four o'clock in the afternoon they were about to tackle the final post hole. Both men were showing the effects of a strenuous day. Their shirts were dark with sweat and their faces were even blacker with dirt and dust.

The metal tailgate was showing the effects too. It was torn and pitted with holes. It didn't even resemble a tailgate any more. In short, it had had it.

Bert leaned on his shovel and regarded the shredded tailgate thoughtfully.

"What are we going to cover this last charge with?" he said.

Frank paced around with one hand in his pocket and the other on his hip while he considered the situation. All at once he straightened up and his face beamed.

"I've got it! There's an old mattress up in the attic. It's good and thick. That ought to cushion the charge. Come on, let's go fetch it."

So they trudged up the back porch stairs to the attic. The mattress gave off a cloud of dust as they dragged it down the stairway and across the lawn. With a grunt and a sigh they flopped it down on top of the last dynamite charge. They shoveled some dirt on top of it.

"Amen!" said Bert as he straightened up.

The two men retreated around the corner of the barn, and Frank set off the charge.

In what seemed like slow motion, the mattress disintegrated. It exploded in a shower of white gossamer fluff that rose like a mushroom cloud from an atom bomb.

The two men walked out from the corner of the barn and stared with unbelieving eyes upward into a white snowstorm.

"Feathers!" Bert guffawed. He clasped his belly with both hands and roared with laughter. "A feather mattress!"

The feathers were a mixture of duck down and goose down. They drifted ever so slowly, covering the driveway, the barn yard and the lawn with a white film. Many of the wisps of down were so light they refused to descend but just kind of floated around, suspended in the air.

After a minute of stunned silence, Frank uttered an oath. "My Gawd! The wires!"

Sure enough, the service wires that went from the electric

power pole to the barn had been severed by flying rock. How in tarnation would they be able to do the milking? Frank stomped into the house and phoned the power company.

The repair man who responded had the wires spliced and hooked back up in short order, so that Frank was only two hours late for the evening milking. The repair man looked at the carpet of white feathers scattered on the lawn around the hole that had been blasted for the post.

"Who's the explosive expert?" he asked.

Frank was silent, but Bert had about all he could do to keep a straight face. "You're looking at him!"

WINTER

RAY NAGELE

I did not need phone-a-vision to picture the expression on Ray Nagele's face at the other end of the line. Years of cattle buying had conditioned him not to show anything resembling enthusiasm.

He sounded far from enthusiastic about my request for him to truck two of my heifers 150 miles to Stewart Field airport at Newburgh to join an export shipment to Saudi Arabia.

It would be a long, wintry drive on this cold January day. And it was awfully short notice. I phoned him at 10 a.m. And the heifers had to be at the airfield by 5 p.m.

My heifers were supposed to have been picked up the previous day by a trucker from the export agent. But there had been a delay in the return of their blood test results for blue tongue and leukosis, so the heifers missed out on the scheduled truck.

Then this morning I got a phone call from Warren Wigsten, the export agent.

"I just got word from the state veterinarian's office in Albany that the test results came in this morning," he said. "I can still take your heifers if you can find a way to truck them to Newburgh today."

Warren was anxious to get my two heifers to complete his export order. And I could use the extra money the heifers would bring. But time was crucial; the heifers had to be at the airfield before this evening.

One thing was in my favor as I tried to persuade Ray. There hadn't been much snow lately and the roads were bare. Then I mentioned my clinching argument to him.

"The exporter told me he will pay a dollar a mile for trucking."

Without a moment's hesitation Ray said, "What time do you want to leave?"

"How about noon? That will give us five hours to get there."

"Okay, plenty of time. See you at noon."

I phoned Warren and told him I had found a trucker. "Now give me directions to Stewart Field from when we get off the Thruway at Newburgh," I said.

Warren's directions were clear. Then he finished with: "There is a side entrance gate to the airfield; it's near the cattle barn. It will save you a lot of hassle because Stewart Field is a huge place. The gate always is locked before dark, but I will arrange to have my man, Pete, keep it open until five o'clock. Oh, and by the way, on your way down can you stop at the state office building in Albany and pick up the health charts on your two heifers. You can get them from a Miss Dott."

Ray Nagele's red cattle truck backed up to my barn door fifteen minutes before noon. We loaded the two heifers, gave them a deep bed of straw, and started off on schedule.

It was a clear, cold, typical mid-winter day. Ray cruised along at the 55 miles per hour speed limit. In an hour and a half we were in the parking lot of the state agriculture and veterinary building in Albany.

"It should only take me a few minutes to pick up the health papers," I assured Ray.

When I got to the office on the fifth floor, Miss Dott had the health papers ready on her desk.

"They haven't been signed yet," she said. "I'll locate one of the state veterinarians here in the building to sign them for you."

After a short conversation on the phone she replaced the receiver. "Dr. Smith is in another building but will be here in a few minutes."

I sat down in the chair by her desk to wait. As I gazed around the large office, I was impressed by the relaxed family atmosphere. Several secretaries were knitting. At another desk a man was reading a novel. Miss Dott seemed to be the only one who was working.

The minutes crept by — five, fifteen, twenty — and I began to get nervous. It was nearly two o'clock. There were only three hours remaining for Ray and me to drive the remaining 100 miles, and we wanted to arrive at the airfield before dark.

Another thirty minutes dragged by. Just as I was getting

ready to jump to my feet and scream "Where the blazes is he?", Dr. Smith strolled in.

I rose to my feet expectantly, but first he stopped and had a leisurely chat with one of the secretaries. Finally he ambled over to Miss Dott's desk. She handed him the health papers. He flipped through them and affixed his signature.

I grabbed them and dashed for the exit. I ran down the hall and punched the button for the elevator. It was way down at the lowest level. Unable to wait, I ran down the five flights of stairs. As I galloped out to the parking lot I half expected that Ray had given up hope and driven off, but the red truck was still there, the engine running to keep the heater warm.

"What happened?" he asked as I climbed in the seat and slammed the door.

"You'll never believe it. I've been sitting up there all this time waiting for the vet to come and sign the papers."

Ray shifted gears and the truck moved off. "You were in there an hour. A whole hour!"

"My God," I fumed. "I still can't believe it."

In another ten minutes we were back on the Thruway.

At 4:45 we finally got off at the Newburgh exit. By now it was sunset and the quick winter darkness began to fall. We still had nine miles to go to reach the airfield. As the truck sped along, the cold feeling of doom descended on me. We would never make it in time.

Finally the airbase sign appeared in the glare of the head lights. We turned off on to the side road that bordered the base, separated from it by a tall wire fence.

We hurried along, looking for the side gate. It was difficult to see in the glare of headlights from oncoming traffic.

"There it is!" I shouted.

Ray pulled off on the shoulder. The gate was closed and padlocked.

"Now what do we do?" asked Ray.

"Let's turn around and find the main entrance."

Ray backed the truck around and we drove a couple of miles to where we had seen the sign. We turned in, but it proved to be the entrance to the housing area for the military portion of the base.

We drove on, four or five miles further, desolate farmland on one side of the road and the fenced airfield on the other. We passed several gates that were barricaded.

"Would you believe we've driven ten miles since we turned around?" Ray said finally. "This sure is a big airfield."

"We'd better turn around again," I had a sinking feeling as Warren's words came back to me: 'Stewart Field is a huge place and I wouldn't want you to get lost...'.

"There was a gas station back by the military base," I said. "I can phone the cattle barn from there."

Ray found a driveway and backed the truck around. Then we sped back the way we had come.

"Lucky we haven't seen any deer," Ray commented. "They usually come out just as it gets dark. Saw a couple of deer crossing warning signs a ways back."

As he spoke the truck went over a knoll and around a curve. The headlights swept ahead, illuminating three deer just stepping out of the bushes. The deer trotted across the highway a hundred yards before us.

I caught my breath. Ray let up on the gas.

A fourth deer stepped out of the gloom and a fifth deer

followed. They dashed across the highway just inches ahead of the truck.

"Whew!" Ray turned his face to me. "I thought we were going to hit!"

"If there had been one more..." I said.

We drove to the gas station. I made the call to the cattle barn. Pete gave me new directions:

"Go past the military base two blocks to the firehouse. I'll phone the fire patrol and ask somebody to escort you to the cattle barn."

A few minutes later we pulled up beside the firehouse. One of the men there hopped in a pickup truck and scooted off down a deserted blacktop runway that was being repaired. We followed in pursuit, dodging sawhorses, barrels and construction warning signs.

After driving like a streak for two miles, the pickup's taillights flashed red as the driver braked, then did an abrupt U turn and stopped.

We slowed and Ray rolled down his window. The driver of the pickup leaned out.

"This place is so big, I'm lost myself," he said. "We've got to go the other way."

We followed again down another runway and made several turns onto diagonal runways. Then after another couple of miles we arrived at the cattle barns — an isolated cluster of buildings in the middle of nowhere.

"We never would have found it on our own," Ray exclaimed. "We would have driven around all night!"

By now it was 7 p.m. The rest of the story was anti-climax. We led my two heifers off the truck and in to the barn to join the other export heifers in a pen where there was hay in a manger and water. They would have a brief rest before boarding a jumbo jet and continuing their journey overseas.

The five men in the barn crew were cold and hungry from waiting around for us. We all drove back to the village for supper at a diner.

"Thought you told me this morning we'd be home for supper," Ray said, chuckling as he buttered a slice of bread.

"All's well that ends well," I replied, trying not to think of the long 150 return drive still ahead of us.

We made it okay and arrived home not too long past midnight.

THE DEER TRACKER

It was just eleven-thirty in the morning when I turned into the Northrup dooryard. Victor was supposed to ride with me to the farm meeting and I was stopping to pick him up.

I turned the pickup around in the new-fallen snow of the yard and waited with the engine idling. A few minutes went by and Victor still hadn't come out of the house, so I went and rapped on the kitchen door.

His wife, Clara, answered my knock.

"Is Victor ready?" I asked.

She looked worried. "No, he's not. He's not even back yet."

"Back yet? What do you mean?"

"He went tracking a deer about nine o'clock this morning. He said he was sure he would be back for breakfast before ten." She turned a fretful glance over my shoulder at the long, snow-covered hill behind the barn. I turned and followed her gaze, but all I could see was the snowy slope with brown bushes and the bare trunks of trees here and there on the hillside. There was no movement, animal or human, no sign of life of any kind.

We stood there awkwardly while the chilly wind rattled the half-open storm door.

She gave an involuntary shiver. "You'd better come inside. I"m sure he'll be along any minute now."

She poured coffee for me and I sat at the kitchen table while she explained what had happened.

"We had just milked the last two cows," she began, "when Vic looked out of the barn door. 'By God, there's that deer!' he said. And sure enough, a buck was standing in the bushes at the edge of the old apple orchard. It was the same one he saw off and on all summer, and he's been trying to get it ever since hunting season opened two weeks ago.

"He ran and got his rifle and took a shot right from the barn door. About scared me and the cows out of our wits!

"It was a long shot, but the deer went down. Then it got up and took off over the hill. Vic said to me, 'Honey, you finish up here, will you? I'm going to follow that buck and get him. I won't be gone long at all.'"

"So he took right off to track the deer?"

"Yes, and that's more than two hours ago. Lord! I hope he hasn't hurt himself."

I hastened to reassure her. "Now don't worry about Victor. He's all right. That deer is just taking longer to go down than he figured on." I tried not to think of the mishaps that could plague a man alone in the winter woods. Slipping on a fallen tree trunk hidden under a blanket of snow, a man could break an ankle and be nigh near helpless.

It's funny how time crawls when you're waiting for someone. It seemed an eternity that I sat there sipping my coffee, but in reality only about twenty minutes went by. I tried to make small talk, but in the back of my mind the neural circuits were calculating how long it would take me to drive back home, change clothes to hunting pants, parka and heavy boots. I would have to follow his trail and look for him, that's all there was to it.

As I pushed my chair away from the table and got up, there was a sudden thumping on the back stoop. The kitchen door flung open and Victor staggered in. He was a sorry sight. His parka was soggy from wet, melting snow; his pants were soaked from his boots to his knees; and his face was red with exhaustion.

He leaned his rifle against the wall and groaned as he pulled off the frozen slabs that were his mittens. With labored grunts he kicked off his boots and lurched toward the stove, peeling off his parka on the way.

"Gawd!" he croaked. "Give me something hot to drink!" It was more like a strangled moan than a voice.

Clara poured a mug of steaming coffee. She steered him to a chair and helped him peel off his drenched trousers. He did not protest as she lifted one leg and then the other and pulled off the dripping wool socks, clucking disapprovingly at the sight of the livid red skin of his feet.

She filled a big pan with hot water from the kitchen sink and set it by his chair. Victor winced as she pushed his long underwear up to his knees and lifted his feet into the water.

"Serves you right, Daniel Boone!" she said. "Where in the world have you been all morning?"

He pushed his empty mug across the table. She filled it again from the coffee pot. He sucked it down hungrily, and still his teeth chattered as he began to explain what had happened.

"Gawd! I must have walked ten miles!" He sneezed loudly twice, and then again.

"That deer left such a trail of blood that I was sure I would find him dead around the next hill. But he just kept going. I never did see him but once, and even then there were too many trees in the way for a shot."

Wordlessly, he held out the mug for more coffee and drank half of it before continuing his story.

"I tracked him clear to the end of the farm, then across Dan's cornfield and through the swamp. The buck crossed the creek there. That's when I broke through the thin ice and fell in.

"Then I trailed him through a thick grove of hemlock and came out on the west end of the Diefenback farm."

I whistled. "You sure did travel some."

"Gawd, did I ever! I should have turned back then, but that buck had such a rack on him, I couldn't let him get away. Before I knew it I came out on Hardscrabble Road."

I whistled again. "That is darn near ten miles."

"And more, when you think of all the ravines I climbed up and down. I was so plumb tuckered out by then I just couldn't go another step. That's when I met two other hunters coming out of the woods on the other side of the road. I told them if they wanted to take over and track my deer, they could have him — but only if they gave me a lift home first. They said okay."

"So they drove you here and then went back?" I said.

"Yup. And a wild goose chase too. They'll never find that buck. He's probably in the next county by now. My mistake was I never should have got on his trail so soon; I should have come in for breakfast and given the buck a chance to lay down and get stiff."

"You're right," I said.

We didn't hear a vehicle drive in the yard, but suddenly our conversation was interrupted by a knock at the kitchen door. Clara opened it. There stood two hunters clad in red jackets, both of them smiling.

"We just stopped by again to say thank you for the nice buck," one of them said.

"And seeing as you shot him, we want to offer you some of the venison," the other one added.

I stepped closer to the door and looked out. On the back of their pickup lay a magnificent buck.

I turned and looked at Victor and whistled. "A ten-pointer."

Victor shot bolt upright in the chair and craned his neck to look out the window. His eyes popped and his jaw dropped. He stared at the two men and managed to croak out one question.

"Where did you find him?"

"In some bushes not two hundred feet from where you left off. We almost tripped over him. He was deader than a door nail!"

Victor sagged down in the chair, an agonized look on his face. One muffled word escaped from his lips.

"Gawd!" he said.

FROZEN PIPES

I was squatting down in the manger, cutting a short strip of rubber from an old inner-tube when I noticed the cows lift their heads and stare down the feed alley toward the barn door. Someone had entered the barn.

Using the tin shears I finished cutting a rubber patch a couple of inches wide and twice as long, enough to wrap around the water pipe. Then with the screwdriver I opened up the stainless steel clamp.

Footsteps approached along the barn floor, and presently from under the nearest cow's belly I saw a man's legs encased in heavy insulated winter boots. I recognized the big feet even before the voice spoke.

"Hello, Richard. What are you up to?"

It was the feed salesman, Jefferson Dodge.

"I'm patching a small leak in this water pipe," I said, indicating a thin spray, more like a mist, that arched up over the water bowl and then spattered down to make a small puddle on the concrete floor of the manger. The spray came out of a pinhole in the galvanized pipe of the water line.

"You're using the old gum and rubber band technique, I see."

"Right," I replied with a grin. "This will have to hold for now. I haven't got time to take this line apart today, not in this cold weather."

I wrapped the inner-tube patch around the pipe and then, as I tightened down the clamp, the leak stopped. "There, good for another fifty years!"

"Absolutely!" Then Jeff's smile changed to a look of concern. "Say, you know it does feel awfully cold in this barn. Did any of your water bowls freeze during the night?"

"Yes, this whole west side froze. It's a good thing I shut the water off in the barn before I went to bed, or I would have had some burst water pipes this morning."

"It certainly was awfully cold last night," Jeff said. "Up on the hill my thermometer said five below zero, but with the bitter wind the chill factor was something like forty below."

"That wind was terrible; it didn't let up all night long," I said. "Thank God it's starting to go down now because the temperature outside has barely climbed to zero."

I walked over and squinted at the big thermometer hanging

on the wall in front of the cows. First I had to rub off the dust and frost before I could see the red line. "It's only thirty degrees in the barn right now!"

I opened up the wall cabinet and put away the tools. Jeff followed me to the milkhouse where I filled two pails with hot water.

"Now what are you up to?" he asked.

"I've got to thaw out the water line on the west side so the cows can drink. Those cows haven't had any water yet this morning." I hoisted the pails. "Say, grab that sweatshirt hanging there, will you, Jeff? And that wool scarf too." I walked through the swinging door into the cow stable.

Taking the sweatshirt from Jeff, I soaked it in the hot water and wrapped it around the first section of water pipe. After about a minute I tried the paddle in the water bowl close by, and a small trickle of water began to fill the bowl. As soon as they heard the sound of water flowing, the two cows on either side of the bowl tried to drink at the same time, each one shoving the other's head out of the way.

"Hey, girls, one at a time," said Jeff. "Boy, they certainly are thirsty."

I soaked the sweatshirt again, and the scarf, and wrapped them around the next section of pipe. Soon one pail of hot water was gone; I asked Jeff to refill it in the milkhouse again.

It was a slow process, but in half an hour all of the water bowls for the twenty-two cows on the west side had been thawed and those cows were drinking. Next, I thawed the water bowls in the four big calf pens at the rear of the barn, but the pressure was so low from the cows all drinking at once that the calves couldn't get any water; just the merest dribble came in their bowls. They bawled their frustration; the din from sixteen calves was deafening.

"I can't stand that racket," I said. "I'm going to carry a few pails of water from the milkhouse for them to drink. Give me a hand, would you?"

"Gosh, Richard, I'm supposed to be on my way making farm calls," he protested. But he followed me to the milkhouse and we each carried two big pails of water back to the calf pens.

It was a long walk, since the barn is one-hundred-fifty feet long and the milkhouse is at the front while the calf pens are in the rear. We had to make four trips before the thirst of all the calves was quenched.

"You couldn't have picked a better morning to stop," I said to Jeff. I certainly appreciate your help. Say, how about helping me feed these cows their corn silage while we're at it? It's too cold to turn them out to the barnyard feed bunk."

"I don't know, Richard; I'm beginning to wish I hadn't stopped in to take your grain order today!"

But he good-naturedly followed me to the silo room.

"You don't have an ensilage cart?" he said. "What will we use, a wheelbarrow?"

There were some empty five-gallon plastic pails in the corner; they originally had held silage preservative. I held out two pails to Jeff. "No, I don't have a wheelbarrow. We'll use these." I picked up two more pails and packed them full from the pile of silage.

"This must be called high-tech farming." Jeff said as he filled his two pails.

I chuckled. "You named it! This goes with that computerized ration-balancing you did for me!"

"I thought my days of feeding cows with a bucket were long in the past," Jeff said. "This is how we farmed back in Minnesota when I was a boy. I thought I graduated from that when I graduated from college and became a feed salesman!"

"You never know what's in store for you when you stop in at my farm," I said.

"All in the line of duty," he answered.

TWO DOORS FROM GEORGE BUSH

The milk tanker was taking a long time backing up the driveway to the barn. It couldn't be the fresh snowfall causing problems; I had plowed the yard twenty minutes ago.

Halfway in the driveway the rear end of the tanker angled off toward the lawn. The air brakes hissed; the tracker pulled ahead to straighten out, then backed again.

When Larry finally walked in the milkhouse he was shaking his head. "I was about ready to ask you to come and back it up for me," he said. "Never had so much trouble. I guess it's just one of those days."

"For a minute I thought it was Fritz backing in," I said, grinning.

Fritz had been a relief driver the previous months, and the mention of his name was guaranteed to bring a chuckle. His prior experience had been as an over-the-road truck driver, hauling milk tankers downstate to New York City. Straight highways were what he was used to; narrow farm drives gave him no end of trouble.

When our milk route's regular relief driver went on extended sick leave, Fritz was tapped as a replacement for a couple of months. During his first week as replacement he rode with the regular drivers in order to learn the various farm routes. I remember the morning after the day he rode with Larry.

"I don't know whether to believe that guy, he's got such a line of bull," Larry said as he hooked the hose up to the milk tank outlet. "Fritz rode with me all day yesterday and he just about talked my ear off. The first thing he told me was that he spent thirteen years in the Marines. Then he said that after that hitch he was in Special Forces, I forget how many years, and it ruined his marriage because they were always calling him at weird hours of the day or night and he had to go. Then he said he went from that job to the C.I.A., but he had to give that up because it got so he couldn't trust people."

"How do you mean?"

"The way he explained it, his job was to dig up evidence on espionage on various people, and after a while it got so he could never be sure who was out to get revenge on him."

Larry flicked the switch and the pump started sucking milk out of the tank.

"Well," Larry continued, "this went on all morning while

he was riding in the truck with me — him talking a blue streak. About noon the tanker was full, and we had to run it down to Walton and drop it off at the cheese plant.

"Walton — that's up in the hills, the Catskills. We were barreling along a winding road through the mountains, and Fritz mentioned kind of off-hand that the State Conservation Department released twelve pairs of bald eagles there. Right then I knew what was coming next.

"Sure enough, in the next breath he told me that he himself released one pair. He said he trucked all twelve pair all the way from California — just three-hundred pounds of eagles in a big tractor-trailer rig. He said he got so attached to them, he asked the conservation guys if he could release one pair of eagles himself. So they let him — according to him, anyway.

"Before the two birds flew away, he said they hopped up – – one on each of his shoulders. Then they flew up high and circled once. Then they swooped down, almost taking his hat off, as though to say goodbye."

"Quite a story," I said, laughing.

During the following month, Fritz was at my farm twice a week because that was the schedule: Larry drove four days straight, and then had two days off. I didn't get to meet Fritz the first couple of weeks because he was always more than an hour late.

Eventually the day came when Fritz arrived on schedule and I got a chance to meet this James Bond character.

I expected a former Marine and C.I.A. agent to be an imposing figure; but the person stomping the snow from his boots as he came in the milkhouse was an ordinary-looking man of average height and medium build. I had expected someone with the hulk of a fullback. Could this be the man who had done everything and been everywhere?

He went calmly about his work; he lifted the cover off the tank; he read the milk weight off the dip stick; he took the sample of milk.

A trifle disappointed, I turned back to the sink where I was washing up the utensils. The radio on top of the milkhouse hot water heater purred out the day's weather forecast, and then blared out a news item about an avalanche in Switzerland that had taken the lives of two American skiers.

"Rugged country, those Alps," Fritz remarked.

"Have you been there?" I asked.

"Yeah, once. Did some training with the Swiss Guards."

Mild surprise must have shown in my face and in my voice. "Is that right?" I said.

I waited for him to elaborate, but when he didn't respond, I thought some prompting would encourage him.

"You must have seen some things in your days," I said.

He gave me a knowing look.

"Oh, I wasn't always a dumb truck driver. At one time I was only two doors from George Bush."

I didn't know what to say.

LOREN and NELLIE

The snow was tapering off when I got in the pickup and drove up the hill on the side road on my way to feed the heifers at the upper barn. A lazy shower of flakes sifted through the afternoon sunshine. The town snowplow hadn't been through yet, and the country road was an unrumpled blanket of white goosedown.

December days are short; at 3 o'clock the sun already was taking a westerly slant toward the woodlot on the hill above the creek. I drove slowly, admiring the beauty of the scene. The perfect whiteness of the fresh snow was now beginning to be tinged with pink reflected from the rose-colored clouds the sun was dragging with it to the hill. In another hour there promised to be a spectacular winter sunset.

As I approached the driveway to the heifer barn I could see, further up the road, neighbor Loren standing beside the fence to his horse pasture; he seemed to be swinging a sledgehammer. Curious, I steered the truck up there. He was trying to drive a fence post in the frozen ground.

Standing in the open gate close by him was his wife, Nellie, in fur-lined boots and a coat to her ankles. Keeping a close eye on the two horses in the paddock — a palomino quarterhorse gelding and a brown Morgan mare — she gave her husband encouraging advice as he swung the maul.

I brought my truck to a stop and rolled down the window. "Horses get out?" I asked. An unnecessary question because it was plain to see what had happened.

Nellie turned to answer me, and the horses, alert for just such an unguarded moment, made a quick feint to dash around her through the gate, but she flung out her arms just in time.

"Oh no you don't!" she sputtered. Then she glanced at her husband. "Loren, if that post won't drive in deep enough, for heaven's sake, get a skinnier one."

Loren set the post maul down as he took off his wool cap and wiped the sweat from his forehead. "Horses make divorces," he said, with a wink in my direction.

I chuckled to myself. They were both in their 70's, and after fifty years of marriage there wasn't a more devoted couple. Of course, a sense of humor kept them lively. Horses make divorces, indeed!

Loren replaced his wool cap. Turning back to the gate post,

he gave it a solid whack with the maul. Before he swung the sledge again, he gave another wink and said, "All I want when I die and go to heaven is horses that don't break out in the middle of the night."

Whack! The post settled another inch.

Now it was Nellie's turn. "Who says you're going to heaven?"

Another whack with the maul; the post definitely was going deeper. "Well, let's just say — when I get to the other side." Whack!.

I just sat there in the truck and smiled. There weren't many people like them left in this part of the country. Retired more than ten years now, they had farmed all their life together. By a stroke of good fortune, just when they were thinking of retiring, they had been able to sell their dairy herd for a real good price.

Although the barn was now empty, Loren and Nellie kept the farm because they hated to move out of the comfortable home where they had spent their entire married days, but he often said later, "When the cows go, the farm should go too. It's too hard keeping up with the taxes when there's no milk check."

They managed though, by selling the hay in the summer and raising a few calves every year to sell as bred heifers. And, as their one luxury, they kept these two old saddle horses.

They had always liked horses and riding. Throughout the good weather of spring, and right in into summer and fall, their favorite pastime was to saddle up and ride the abandoned country roads and old logging trails. Even in winter, a week seldom went by that I didn't see them trotting past my farm driveway for an afternoon outing.

Loren gave the post a final whack, and even Nellie was satisfied. I got out of the truck to help hoist the gate on the hinges, and held the fence planks while he nailed them fast. The sun was now angling deeper in the southwest.

"Time for coffee," Nellie said. "Come and have a cup with us."

I accepted with alacrity, first helping Loren carry the maul, crowbar, shovel and the rest of his tools to the tool shed. I marveled, as I always did, at the neat arrangement of his workbench. Wrenches and chisels and pliers were in orderly rows on a wall rack. Nuts and bolts and nails and washers were in plastic jars. All the shop lacked was a sign reading, 'A place for everything, and everything in its place.'

Orderliness and doing things right were the keystones of Loren's life. He often remarked that you could judge the condition of the entire farm by the way the manure spreader looked. "If the manure spreader is filthy and beat up, you can pretty well guess the rest of the farm looks that way too."

The pungent aroma of coffee greeted us in the kitchen. And there was a fresh batch of oatmeal cookies. I took off my heavy winter jacket and sat down. Loren was still in a jovial mood. He took a long swig of coffee from his cup and settled back in the kitchen chair.

"Yessir, when I get to the other side, I hope I don't find any horses there," he said, winking again. "My idea of heaven is no chores to do, and nothing that breaks down, or fences to fix."

Nellie smiled and passed the plate of cookies. "Well, just in case, you'd better bring tools," she said.

After a second cup of coffee, I reluctantly pushed back my chair; it was time for me to go and do my chores. The rosy ball of the sun was now low in the sky, and the troop of pink clouds was hurrying to meet it.

"By golly, it's going to be a beautiful sunset," Loren said to Nellie. "Put on your parka. Let's saddle up the horses and ride up the hill to watch it!"

SWAMPY WALRATH

Whenever I drive on the Vedders Corners road and go up the hill past the old farm that adjoins ours, I think of Dwight Walrath. Known affectionately throughout the countryside as 'Old Swampy Walrath', he was my father's neighbor for years and years.

A long time ago, before that section of the county highway was paved with macadam, the road up the hill was gravel. One spring day Swampy drove his team and stoneboat down that hill to help my Dad pick stones in a plowed field being fitted for planting. The loose gravel under the smooth planks of the stoneboat acted like ball bearings. The stoneboat got rolling on the gravel and bumped the hind feet of the horses, scaring them.

Swampy pulled back on the reins, but this only seemed to make the stoneboat roll faster, bumping the horses' feet again and again. In no time at all Swampy was hanging on the reins for dear life with a runaway team.

He managed to stand upright only a short way as the stoneboat plummeted down the hill. Soon he lost his balance and fell, but kept a fierce grip on the reins as the team dragged him along, half out of the stoneboat.

At the bottom of the hill the road leveled off, but the terrified horses kept galloping. Swampy was dragged another half-mile before he brought the team under control and to a halt. He was battered and bruised from head to foot from the ordeal, but indomitable.

A lesser man would have let loose of the reins, but not Swampy. He feared what might happen if he did let go and roll off to save himself: the uncontrolled horses smashing the stoneboat to splinters and destroying themselves along with it.

When Swampy eventually retired from milking cows, he sold the farm to a family that had a dairy there for about ten years, until the owner had the misfortune of being gored by a bull. There hasn't been a dairy on the place since then.

The next owner was an absentee landlord who initiated the decline of the farm by selling off the house, which stood across the road from the barn, as a separate residence. Then he let the loggers into the big woodlot.

After the woodlot was ravished, the farm was sold again, this time to an absentee landlord from Long Island. He let a neighboring young farmer cut the hay and work the cropland for a few years. But during the farm recession of the 1980's, double-digit inflation ruined that heavily-mortgaged young farmer; the bank foreclosed; the young man went bankrupt and Swampy's farm has been idle ever since.

It's funny how fast a barn can run downhill when there's no one to look after it day by day. No one to mend the loose window pane that rattles in the wind and eventually falls out. No one to keep the barn siding nailed on tight, or to notice the loose sheet of roofing tin peeled back by the storm; no one to keep the pigeons out of the haymow.

Every spring the weeds grow taller in the barnyard. The lush burdock shoulders its elephant ear-sized leaves high above the window sills of the stable. Blackberry vines creep in a small jungle from the old machinery shed right up to the haymow ramp.

What the place is longing for is someone to take it over and cherish it. When I pass by I almost can hear the barn call out, "Sakes! Won't someone keep a dairy here, or at least raise some heifers to make my stable cozy again? I need someone to sweep the cobwebs off my beams, point up my loose foundation stones with mortar, and even give the clapboards a coat of new paint."

I don't suppose the farm will ever support a dairy herd again. It's barely a hundred acres, and twenty of that is woods. The hayfields are all sidehill, with many wet spots that would require tile for drainage. The only flat land is on the north side of the barn, and that land is also swampy and poorly drained.

Which reminds me to explain how Old Swampy got his nickname. You see, in addition to milking cows, Swampy had a sugarbush and made maple syrup in springtime. When the roads got passable in early March, he drove his bobsled and team the three miles to the village to recruit help for sugaring off. On the way he passed through the 'Hollow' along Klock's Hill and exchanged good-natured bantering with Seeber Crouse and the other boys who lived there.

"Hey there, you muskrats in the hollow," Swampy called out. "Why don't you come up to the farm and help gather sap. I'm going to be sugaring off this week."

"Why it's the old swamp toad! Is it maple syrup time again?"

And Dwight would have half a dozen lads climbing aboard the bobsled on his return trip. The old swamp toad eventually became 'Old Swampy', and the name stuck to him to his dying day.

Even though the farm changed hands several times since Dwight retired, it is still known as Swampy Walrath's place. When a man becomes a legend, his name clings for a long to the place he loved.

We may never see his like again: a big hearty man of prodigious strength who could lift a steel railroad rail all by himself; a man who never drank, or swore, or smoked or chewed tobacco.

If he had a vice, it was that he liked to play cards. Once or twice a month he was in attendance at an all-night poker game in the village. He was good at winning too, stuffing the folded up bills in the small pockets of his vest.

After the game was over he jogged his team home in the early morning hours, just in time for morning chores. He always went into the kitchen first to show his winnings to his wife. As he took the crumpled bills out of his vest pockets, one by one, and unfolded them on the kitchen table he grinned proudly.

"Lookee here, Liz, look at that bill! And look at that one! And here's another!"

Then he'd have a cup of coffee and head out to the barn.

He was a man strong as a bull, gentle as a kitten, and with the constitution of a war horse. The following incident is still remembered hereabouts:

One winter they were cutting ice on the pond at Vedder's Corners. Swampy had just finished loading his bobsled with cakes of ice when he accidentally trod on a sawed section of the pond and fell through the ice. When they hauled him out of the frigid water, he was thoroughly soaked, from his felt boots clear up his wool pants to his heavy wool jacket.

Undaunted, Swampy clucked to his team, urging them toward home in a steady jog. He himself trotted beside the sled the three miles home, and generated enough body heat to keep from freezing.

"No, I don't suppose the farm will ever see his like again," the old vet, Doc Cairns said. "After they made him, they cracked the mold."

Still, I hope someday someone will come along who will fall in love with that small farm and cherish it the way it yearns to be cherished. Someone who wouldn't mind haying on sidehills, who would rebuild the fences and fill up the wild, overgrown pastures with tame heifers.

All it needs is for someone who loves the quiet country to take a walk over the thawing hayfields on a mild afternoon in March — down where the side hill slopes south to Swampy's sugar bush, with its old sap house still standing among the maples, where the warm sun casts long blue tree shadows on the white snow. It would be love at first sight.

COMPETITION

We sit at the supper table enjoying the only meal we are able to partake of leisurely as a family when the school year is in full swing. Breakfast is a hurried affair that is eaten in haste before the school bus comes along our country road.

"Please pass the applesauce, Dad," says my daughter, Ann.

"How was school today?"

"Yuk. Mrs. Dillenbeck gives us too much homework."

"That is quite a bit of homework. I can't remember having that much when I was in third grade." (She has six pages of reading to do for social studies. They are studying China — the

Yangtze River, the Yellow Plain, irrigation and so forth. And a full page of mathematics, delving into the mysteries of subtraction which so far elude her.)

"We had a fire drill today."

"Oh. Did all the classes get out quickly?"

"No. The principal made us do it over again. All the kids were talking and goofing off the first time. The principal got mad and said that the way our tongues were flapping, if there were a real fire, we never would make it safely outside. He said that during a fire drill the only thing he wanted to hear flapping was the soles of our shoes."

"He's right, you know."

"Yeah, I know."

Silence reigns for a while.

"Did Phoebe have her calf?" she asks.

"No, not yet."

"I hope she has a heifer."

"What did you have in cafeteria today?"

"Spaghetti. Yuk! They get it out of a can."

"Let's see. Was today your day for music, or gym?"

"Gym. Music is Tuesday."

Silence again. The sound of forks on plates, My daughter swirls her mashed potatoes to make a lake for her gravy.

"What I don't like about gym is that they make you try to win against your friends."

My fork stops halfway to my mouth as this bit of wisdom assails me. The fork continues its journey and I munch a mouthful thoughtfully.

"What do you do in gym now?"

"Oh, we have calisthenics, and then we play kick-ball."

The happy carefree days of first and second grade are gone. Gym for those two grades was merely running around outside letting off steam. With third grade comes the introduction to organized games — and competition.

The remainder of our meal is eaten with intermittent conversation. My mind keeps returning to that one revealing observation. It's surprising what truthfulness you hear if only you take time to listen. Win against your friends, indeed!

Wherefore this need for competition? I can't remember having thought about it seriously before. Can competition be instinctual? It hardly seems so, if an eight-year-old resents having it instilled in her — with the same resentment she feels against the stubborn mysteries of subtraction.

It's not winning or losing that matters, but how you play the game. At least that's what adults tell each other. But we encourage competition from childhood up, not only in sports, but in grades as well. We're always looking for A's and B's on the report card. Is something wrong with our value system? Learning should be a voyage of discovery with classmates, not organized competition.

Now that my daughter has brought me up short in my tracks, how much competition do I myself live with? I have to admit there is plenty, even with milking cows. For instance, I am a member of the Dairy Herd Improvement Association and each month the rolling herd averages are published in our county extension newspaper. This list is sort of a Who's Who in herd averages. There surely is competition to see who can be in top ten farms on the list, and who owns the top five cows every month for daily milk weight and high butterfat and protein. Are we improving the breed, or improving egos of dairymen?

A certain amount of striving is healthy and normal. A person strives to do better than he did yesterday. Or does he strive to do better than his friends?

This competition to 'win against your friends' is carried through the community, and right up to national and international levels. We arrive, therefore, at the ludicrous situation of 'friendly' countries seeking to outdo each other in the sophistication of their atomic armaments.

As if to underline my thoughts, the radio newscaster comments on the latest U.S. missiles; they ride on underground tracks and are moved constantly to any one of eight different positions so that the enemy will have a hard time guessing where they are located at any particular time — childrens' games in earnest.

What would happen if all the children of the world rebelled against being taught to 'win against your friends"? I have a suspicion the world would be a better place. Out of the mouths of babes, indeed

TINY SHUSTER

Dick Kelly was still shaking his head in disbelief when he pulled his car to a stop in my farm driveway. Dick represented the national Holstein Association as a field consultant for eastern New York State.

As a consultant he was well trained in using all the sophisticated modern procedures for breeding better dairy cattle: Predicted Difference for milk and type, and all the genetic information available from the computer at the Brattleboro, Vermont headquarters of the association. His job was to assist dairymen who asked for advice in selecting sires to use in breeding their cows. The goal was improving functional body conformation, as well as increasing milk production, in the offspring.

When a cow had physical traits that needed improving — a pendulous udder for example — an 'udder improving' bull could be selected to correct that fault. When his semen was used to breed the cow, the resulting calf would grow up to have an udder with strong supporting ligaments that would hold the udder, and its teats, close to the body out of harm's way. No more teats dangling in the mud, or stepped on by a cow's own clumsy feet.

The same method was used to increase milk production in the next generation. The Association had a list of all the bulls in the country who had daughters with official records for the pounds of milk they produced; those records were compared with their mothers' records. If the daughters of a certain bull produced more milk than their mothers, then that bull got an official rating of being 'plus for milk'. It was all very up-to date and scientific, using statistical methods such as 'standard deviation', 'probability' and so forth.

But nothing in Dick Kelly's training had prepared him for his encounter with Tiny Shuster.

"I can't believe it," he said to me. "That woman is breeding her cows by astrology!"

I chuckled. "That's a new one on you, isn't it?"

"She meant it too." His brow furrowed in puzzlement, and it was evident he was still trying to comprehend what he had seen. "She had all the zodiac charts and cow pedigrees spread out on the kitchen table." He looked searchingly at me. "Was she pulling my leg or is she for real?"

"She's for real. You may think she's eccentric, but don't be fooled. She's got a real head on her shoulders. Were you able to give her any advice on breeding?"

"A little," he said.

When Kelly drove away he was still shaking his head and mumbling, "Astrology!"

Over the next few days, the more I thought about it the more I was intrigued by the novel approach of applying astrology to cattle breeding. I was not surprised to hear that Tiny Shuster was experimenting with it. If being first to embrace new ideas is the definition of a radical, then Melanie Jeanine, better known as 'Tiny', is a radical.

Twenty years ago when she and her husband bought the Hyde farm and began dairying, she first displayed her genius for original thinking. She organized twenty of the local farms into a cooperative venture to purchase farm supplies at substantial savings. The approach is commonplace nowadays, but twenty years ago it was cutting edge.

She started out by getting them together to order grain in 200-ton lots, cash on delivery; then went shopping to each of the big feed companies for the best price. This group bargaining power gained them twenty to thirty dollars a ton below the price they would have gotten separately. The same approach was used in buying seed, fertilizer and other supplies.

Her next venture was an attempt to organize a bloc of milk and truck it into the Boston market which twenty years ago was more than a dollar a hundred-weight above the New York market price. That venture almost succeeded but was blackmailed in a series of cloak-and-dagger episodes, including the suspicious burning of a trucker's house. Rumor at the time pointed a finger at the Mafia which some said had infiltrated the milk handling industry. So if Tiny was studying the compatibility of astrology and dairy cattle breeding, I wanted to find out more about it. I talked with her on the phone and then paid a visit to the Shuster farm on Kennedy Road.

Tiny was delighted to spread out her books on the kitchen table — planet position Ephemeris and herdbook — and explain what she had done so far.

"Most people think of astrology as crystal ball gazing and witchcraft," she said. "It's not that at all. It's mathematical and precise, using logarithms to calculate planet positions. Astrology is older than religion. I wanted to see if it could be applied to cattle as well as people."

She flipped through the pages of the herd book. "I've begun by making sun charts of each cow in the herd.

"A sun chart is a circle which is divided into twelve segments or 'houses', like twelve slices of pie. The positions of the planets in each house at the individual's date of birth is a guide to that individual's future life direction."

I interrupted her. "To me that is a basic flaw of astrology," I said. "Granted the heavenly bodies can exert forces here on earth, I would be more inclined to believe those forces would be influential at the time of conception or even immediately before. For instance, those forces might influence the recombination of genes that goes on during the shuffling process of meiosis."

"The only problem," she replied, "is that we don't know the exact time of conception. The hour of birth, as a traumatic time, always has been used as the focal point of astrology."

Tiny stopped at one cow's chart. "For example, in this chart in the sixth house, which is health, we have Saturn which means longevity. This cow should live to a ripe old age. And here is Jupiter in the fifth house, the house of love given; this cow will have many offspring.

"Here is another sun chart with Saturn on Pluto in the sixth house. That means sickness and breeding problems. This cow, in fact, had to be shipped for beef."

She continued leafing through the herd book. "Here is another cow born at the end of January. She is an Aquarius. That's a water sign and that cow should be a 'wet' one — a good milker. She actually is one of our high producers."

"Where do you go from here?" I asked.

"I can't draw any conclusions until I keep the sun charts and corresponding cow records for at least two years. Then I can tell if, for instance, those cows with Pluto in the sixth house do have more sickness; or if the Scorpio cows all turn out to be high-strung and mean; or if the Aries cows are unruly and headstrong; or if the Virgo cows tend to be clean, and so forth."

"If all this proves to be true, you may revolutionize the science of cattle breeding!" I said.

"Who knows? For instance, a big concern in the Holstein world is breeding for strong legs and feet. Aquarius influences the legs and Pisces controls the feet. I hope to plan matings so that cows born with favorable planet positions in these two signs will have good legs and feet."

"My God. That will take no end of calculations."

"Yes. But if I can prove it works, it can be computerized."

Before I left, Tiny opened her planet Ephemeris and calculated a sun chart for one of my cows. I went away with my head buzzing with astrological terms: Venus and the moon into Capricorn in the first house; Saturn in Pisces with a solar eclipse at seven degrees Pisces; Mars in conjunction with..."

I'm going to have sun charts done on the heifer calves born this winter, then follow up when they enter the milking herd two years hence and see how many of her predictions hold water.

Who knows what the future holds? Besides using the sire summaries, someday we may program the computer to pick a breeding date which will ensure a calving date having fortuitous planet positions in the proper houses of the zodiac.

But then I'm forgetting one small detail. It all depends on the fertility of the cow. In order for this to work, the cow has to get pregnant at the hour forecast in the charts.

This dairy business never gets dull.

COYOTE

While waiting at the barn for the veterinarian to arrive and vaccinate some calves this afternoon, I went outside and soaked up some February sunshine. It was so warm — almost forty-five degrees by the old thermometer in the shade of the milkhouse — that I sat on the concrete step by the barn door and treated myself to the luxury of doing absolutely nothing for what amounted to about an hour. Following a frigid night in the teens, the warm winter sunshine was a blessing.

There was not the slightest breeze; the air was absolutely calm. I took off my winter parka and leaned back against the clapboards, feeling the sun's rays full on my face. From the open door behind me, quiet reigned. Having finished a full morning's feeding of silage and hay, the cows were all lying down in their stalls chewing their cuds. The only sound was their heavy breathing.

In the paddock off to the left, the four sheep were recumbent on the snow, their jaws moving rhythmically in rumination. Only rain or sleet makes them seek the shelter of the shed. Even the fiercest snowstorm finds them bedded down outside on the snow. No doubt they are well insulated, the four pregnant ewes being all wool and a yard wide.

A casual glance has the sheep all looking alike. A closer inspection reveals individuality. One ewe has the benign look of a British judge in robes, wavy wig and all. The face of another, due to curl of lips and nose, has a perpetual silly grin.

I was reminded of sculptor Henry Moore and his *Henry Moore's Sheep Sketchbook*. The window of his studio in Sussex, England looked out on a sheep pasture. The sheep fascinated him and he sketched them in idle moments. The result was a delightful book illustrating sheep in all poses — grazing, strolling, butting heads, or just staring off into space.

The snow on my house roof and the barn roof was beginning to melt. Water dripping from the eaves barely broke the silence. Drip, drip, patter, patter. When the air is calm, what a pleasant noise water makes dripping from a roof to the damp ground. I watched a bead of water descend an icicle, gather in a fat drop at the tip, break away and splatter on the ground with a tiny splash.

I listened and tried to clear my head of all thoughts. For that brief hour, I concentrated on being still and simply observing, letting my eyes see and my ears hear. Forcing myself to "have all my senses." Taking in all sensations like a recording device.

I didn't realize what a God-awful business it is simply paying attention. My mind kept wandering off on tangents. It resisted the discipline of concentration obstinately.

I looked at the hillside thawing in the sun and began thinking of what damage the January ice might have done to the new seeding of alfalfa. I had to pull my mind off that track — slap its wrist so to speak — and simply observe and listen.

The barn cat came to sit on the concrete step beside me and began grooming herself. Starting with the left foreleg, she licked all the fur to her satisfaction, then groomed the fur on her flank and belly and hind leg. Then she licked the fur on the other side.

Cats must have a tremendous resistance to infection. If humans had to cleanse their skin in that fashion, they would be ill constantly from ingesting germs. There, I did it again — began thinking. I had to force myself to rest and to simply soak up sensations, like my face was soaking up the sun's rays.

I looked at the row of maple trees that bordered the driveway. Stalwart, they lifted their boughs to the sky. At the tip of each tiny twig the buds were beginning to swell. On a day like this I could almost hear the maples pumping sap.

A movement on an outermost branch caught my eye. A gray squirrel was busy at something there. He must have been nibbling buds, trying to get some greens in his diet.

On the west side of the house, in the old lilac bush, the chickadees were busy at the sunflower seeds in the feeder. A blue jay swooped in screaming, grabbed a beakful of seeds and departed. In a few minutes he was back for more. In the field across the stonewall fence, the four horses were browsing at what was left of hay that I scattered on the snow for them in the morning. With their thick growth of winter hair, they looked like cinnamon bears just out of hibernation.

The horses lifted their heads and looked idly at a furry shape moving like a shadow among the tall brown weeds poking through the snow. It was the coyote, mouse-hunting for his afternoon meal. He has been a visitor to the meadow all winter long, and the horses now gave him no more than a passing glance. His long tail tipped with black trailing behind him, he stepped slowly, almost on tip-toe. He stopped and stared hard at a tuft of weeds, his sharp nose and pointed ears all alert. Then he pounced like a cat, and came up nibbling and swallowing a field mouse.

He continued stalking, and his path brought him so close to the horses that if they stretched their necks they could have almost touched him; yet they paid him no heed. They knew he had his mind on other things.

The coyote picked his feet up carefully and placed them down cautiously like a hunting cat. Stop, stiffen, and pounce. A snap of the lightning-quick jaws, and another mouse was limp in death.

The coyote had a regular schedule. About three o'clock in the afternoon the pangs of hunger moved him out on his rounds. He and his mate had a den somewhere up on the hillside. Last night just as the full moon rose, and I had switched off the barn lights after evening milking and shut the barn door behind me, an eerie sound from the darkness made my skin prickle. The coyotes were serenading from the hillside. I'm sure there were only two of them, but together they sounded like a whole pack. Yip, yip, eeee, owwww, ahhh, yap, yap, yip. What a sound!

The coyote continued up the field beyond my line of vision. Now, from the woodlot, a new sound clamored for my attention. Rat-a-tat-tat, rat-a-tat-tat, echoing from a tree trunk. The pileated woodpecker was sounding taps for winter.

FIELDS AFAR

SOUTHERN CROSS

I had to pinch myself to believe it was real. We actually were on Pan Am flight 812, and those emerald green hills rushing up from the blue ocean to meet the descending jumbo jet really were the green hills of New Zealand.

This long-dreamed-for vacation at last was coming true, and all because of a letter from a farm family in New Zealand. A year ago the Gunthers had read one of my 'Jottings' in which I described making maple syrup on our farm. Annette Gunther wrote and asked questions about tapping maple trees; she closed her letter with an invitation : "If you are ever in New Zealand, we would love to have you visit our farm." That was the spur I needed; it was now or never! It took a full year to finalize preparations and arrange for someone to look after the farm, but at last we had passports, the plane tickets were in hand, and we actually were over the Pacific Ocean.

Just twenty-four hours earlier it had seemed our vacation trip would be stymied at the last moment that April morning. Everything was in order, or so it seemed. Judy and Tim would be in charge of the dairy herd during the two weeks of my absence. I had given them all the last-minute instructions about feeding the calves, the one cow that had been treated for mastitis, and the old cow due to freshen that might come down with milk fever.

We stood there in the manger aisle, two hours before my departure time, when it happened. As though on cue, one of the cows broke her water bowl, and a geyser of water gushed from the broken pipe.

My first impulse was to tear out my hair and scream. But calm prevailed. Tim and I managed to repair the mess; I got myself cleaned up, finished packing the suitcases, and drove off.

Now here we were, my eleven-year-old daughter and I, after twenty hours of flying time. After the confusion of leaping abruptly from Tuesday to Wednesday as the plane crossed the International Date Line, we were rolling to a stop at Auckland airport.

The New Zealand experience began even before we disembarked from the plane. All passengers were requested to remain seated while a team of four smiling men came aboard with aerosol cans and sprayed the interior of the aircraft. It was a precaution against the introduction of unwanted insect pests to that island paradise. In a few minutes we stepped on the ground in New Zealand and were whisked through the friendliest customs entry I've ever experienced.

My original plan had been to rent a car at Auckland airport and drive leisurely the length and breadth of both the North and South Islands, each about 500 miles long and 150 miles wide. But those plans had been made for a three-week trip. Now that our holiday had been scaled down to just two weeks, we only had time to explore the South Island.

Accordingly, we boarded a domestic Air New Zealand plane for the two-hour flight from Auckland to Christchurch, 500 miles south. There we rented a compact car at $33 a day, U.S.

We had left New York in springtime; April snow was melting. In New Zealand we entered autumn. Green fields dotted with white sheep stretched to the horizon. Tall Lombardy poplar trees lifted blazing autumn gold foliage to a blue sky.

The interior of the South Island between Christchurch and Queenstown reminded me of the high plains of our American Midwest. It was seemingly endless flat country speckled with sheep and beef cattle and an occasional pole shed bursting with hay. On the western horizon were the purple outlines of the range of mountains known as The Southern Alps.

The people of New Zealand call themselves 'Kiwis' after their national bird. Kiwi is the Maori name for this wingless bird, a shy, retiring creature of the native bush. However, New Zealanders are far from shy; they are the friendliest people on the globe.

One of our stops was Mt. Cook — snow-capped and serene — with the Franz Joseph glacier stretching its river of ice down the western slope to the Tasman Sea.

A two-hour drive south of Mt. Cook brought us to the southern lakes district and the city of Queenstown. It is nestled in a beautiful bay of Lake Wakatipu on the site of a sheep-shearing station established by W.G. Rees in 1860. Even in New Zealand, noted for its stunning scenery, Queenstown is a jewel, evoking images of San Francisco and Switzerland.

Across the bay is a magnificent range of mountain peaks aptly named 'The Remarkables.' Lake Wakatipu stretches more than fifty miles long, with numerous bays and inlets fingering the mountainous shoreline. We took a half-day cruise on a motor launch to Cecil Peak Station, a sheep ranch six miles down the lake. It's accessible only by boat or light aircraft.

Cecil Peak Station consists of 33,000 acres of alpine grazing land, with 7,500 sheep and a beef herd of 400 Herefords. Along with eight other tourists, we ate a sumptuous High Country morning tea, viewed a short film on the New Zealand wool industry, and went to the sheep shed to watch sheep being shorn.

After leaving Queenstown, we continued on the road south. In Southland we drove through some of the finest pastoral country in the world — lush green pastures stretching to the horizon, innumerable flocks of sheep and herds of dairy cattle grazing green meadows. At last we arrived in the city of Invercargill. In her letter, Mrs. Gunther had advised that when I got to Invercargill I should phone; they would come and meet us and show us the way to their farm. I found a phone booth and dialed the number of Spring Terrace Jersey Farm. A woman answered.

"Annette?" I said. "It's Richard Triumpho. We made it at last. We're here!"

SPRING TERRACE FARM

A land of pure mountain air, snowy alpine peaks, flower-strewn meadows and sparkling streams.

A fisherman's dream vacation spot, where an afternoon's fly-casting can net you your limit of brown trout weighing from four-to-fourteen pounds.

A charming island where no one is more than an hour and a half drive from the seashore with its rocky bluffs, sandy beaches and friendly dolphins sporting in the waves.

A country with some of the finest natural grassland in the world, upon which graze vast flocks of sheep and herds of cattle, and each bend in the road offers unending views of pastoral beauty, peace and tranquillity.

Are those descriptions of four separate countries in a travel brochure? No. All of those things can be found in New Zealand, a paradise as far as I am concerned. I wish I had discovered it forty years ago! It's a country where every experience "begins in delight and ends in wisdom." Yet I never dreamed I'd ever get there.

It's funny how things work out. A year or so ago in one of these 'Jottings' I wrote about the springtime ritual of tapping trees and making maple syrup here on our farm in March. Although I know Hoard's Dairyman magazine is subscribed to by dairy farm families all over the world, still I was pleasantly surprised to get a letter postmarked 'Invercargill, New Zealand.' The letter was from Annette Gunther; she and her husband, Gerald, live at Spring Terrace Jersey farm, at the southernmost tip of the South Island; and she wanted to know more about maple syrup, which the Gunther farm family had never tasted.

Annette ended her letter saying that if ever I traveled 'down under' they would love to have me visit them. So, a year later, wonder of wonders, here I was at Spring Terrace.

That particular afternoon I was standing in the yard if Spring Terrace Jersey Farm, looking at the horizon and wondering which way was north, when Gerald said, "Here comes the mob, full of good tucker."

I turned around quickly, expecting I don't know quite what, perhaps a gang of hoodlums that had been drinking. But all I saw was a herd of lovely fawn-colored Jersey cows ambling down the lane toward the milking shed. Behind them, barely moving on a trail bike, was John, a 20-year-old trainee from Wales.

Reflecting on the term 'mob,' I decided it was an apt description of a herd of cows. Goodness knows that at various times and ways a group of cows can be as unruly as a mob. In fact, I like that term better than our 'herd.'

'Tucker' took a little more thinking for me to figure out, but eventually it sounded plausible. After all, when cows are eating they are 'tucking' all that feed into their stomachs.

As I stood and watched the milking operation that first night in New Zealand, I was impressed with the simplicity and efficiency of it all. The milking 'shed' is what we Yanks call a milk 'parlor.' (And as long as we are into terminology, who ever dreamed up 'parlor'? A parlor it ain't.)

The milking shed was a pole barn enclosed on three sides and open on the side facing the paved holding area. The mob stood chewing their cuds, waiting their turn to step forward into the milking line.

Most of the milking routine was the same as mine: cleanse the teats with a paper towel, wait a few seconds for let-down, attach the teat cups, remove the machine when milk flow stops, dip teats with a sanitizer.

Two steps in the routine were vastly different from mine. The first was the grain feeding. When the cow entered the stall, John gave a couple of tugs on the rope to dump a pound or two of mash type grain in the feed bowl. That's the limit. All their milk is produced on that wonderful free grass. When I told Gerald that I feed two and a half tons of grain to each of my cows per year, we came up with some calculations that gave an interesting comparison. His whole herd of 120 cows doesn't eat as much grain in one month as one of my cows eats during her whole lactation.

The second step in the milking routine that was different was this: Each cow was painted on her flank with a swab of clear fluid. I felt I knew what it was but asked anyway to make sure.

"It's a bloat preventative," I was told.

The herd — oops, mob — was milked in two strings. The Jerseys and Friesians were kept in separate pastures. Here again we run into terminology. In New Zealand a pasture is called a 'paddock.'

New Zealand dairy farmers are experts in grassland management. Paddocks are fenced into small areas of three to five acres, grazed intensively, and rotated. One morning Gerald took me on a walking tour of part of his 500 acres of flat land, giving me a close-up view of New Zealand pasture management.

Pasture mixes consist of a blend of white clover, perennial ryegrass, orchard grass and some red clover. Typically, a mob will graze a three-acre paddock for twenty-four hours — one day and one night — and then move to an adjoining paddock. The first paddock will be clipped if needed to keep weeds down and will get about thirty days regrowth before the mob is turned back on it. We walked through one such paddock where the lush re-growth of grass and clover was well above my ankles.

I thought of farms back home in New York State that typically have one 'day' pasture and one 'night' pasture; these two pastures lasting a whole herd of fifty or sixty cows all summer long. Back home, things like pasture clipping, top-dressing and strip-grazing are unknown.

Water being an essential part of a dairy cow's diet, I was quick to notice that each paddock had a. water trough with a float valve. When I calculated that there were fifty or more of these three-acre paddocks, I realized the investment in the pasture water system was a sizeable one. Gerald uses plastic pipe laid underground.

"To mark off a new paddock means $300 to $500," he said, "when you take into account fencing, water trough, new gate posts and a new gate."

And that's another thing. New Zealanders have some of the finest fencing I've seen. Straight — and I mean straight — fence lines; pressure-treated posts; and five or six strands of high-tensile wire, with at least one strand electrified. In all of New Zealand I saw only one strand of barbed wire.

"I wouldn't have the damned stuff," Gerald said with disdain. "It's an invention of the devil."

As we walked through another paddock that contained at one corner a dense growth of tall trees and underbrush, Gerald explained that this 'native bush' was used as one of the winter paddocks.

"The cows can get in the trees and have shelter from the wind," he said.

Although they seldom have snow on South Island, the winter weather can be cold and damp and the winter winds can be fierce.

The lane we took back toward the milking shed was similar to the one we had taken on our way out — a hard-packed gravel-like surface. Gerald told me it was 'marl' dredged from the four-foot-deep drainage ditch that flanked the lane. The land at the southern tip of the island being flat and not much above sea level means that drainage ditches need to be used extensively. Each district has a commission whose responsibility it is to keep the ditche s open and weed-free. Since the cow lanes are an essential part of the pasture management system, it is essential for them to be hard-packed, well-drained and free from mud in order to prevent hoof problems.

When we arrived back at the milking shed, the last of the Friesians was heading out the lane to the pasture paddock and John was finishing up hosing the holding area while the cleaned-in-place washing of the milk pipeline was just ending its cycle. We walked to the house together and were greeted by a warm blaze of wood in the fireplace of the spacious farm kitchen.

Breakfast was waiting. Annette had bacon and eggs for us all, hot off the griddle. And to go with the toast there were mounds of strawberry jam from her strawberry patch in the back garden. And for the coffee there was a pitcher of that rich Jersey milk, from cows grazing the green pastures of the southernmost dairy farm in the world.

WALES

As dawn diffused the eastern sky with a pink glow and mist rose from the hedgerow-bordered pastures, sounds of awakening day issued from the vicinity of the barn.

The barn was on the opposite side of the courtyard, just a stone's throw from the house where we stopped to spend the night as guests of the Blackwell family at their farm Bed and Breakfast.

There was the muffled padding of hoofed feet from cows in the barnyard holding pen; the distant hum of the vacuum pump; the subdued metallic clank of a stanchion.

Familiar sounds in a strange land are comforting, I thought, as I looked down on the scene from the bedroom window. Mr. Blackwell and his son were milking their fifty British Friesian cows at 'Perth-Y-Pia Farm' near the hamlet of Llanvapley, four miles from Abergavenny in southern Wales.

When I chatted with Mr. Blackwell the previous evening, he mentioned that they milked with a pipeline. Now, as I looked from the bedroom window, I could see that the Blackwells had what we in the States call a California flat parlor.

The climate of southern Wales is mild and rainy, Mr. Blackwell told me, with only a trace of winter snow, and the principal winter forage is grass silage stored in bunkers. The younger Blackwell son had been making the third cutting when my daughter and I arrived the previous evening.

I got up and dressed; then I went outside and strolled around the cobblestone courtyard to take in the early morning sounds and sights of an English farmstead. The dog, a black-and-white border collie, came over to make my acquaintance.

Soon morning milking was done; we joined the Blackwell family at a traditional British hearty breakfast: eggs; bacon and sausage; toast and marmalade; and lots of tea. Mrs. Blackwell insisted we sample her blackberry preserves.

"You must try some on your muffins," she said. "I canned a dozen jars this week, and I do think it turned out quite topping well."

Although I had already stuffed myself from the sumptuous breakfast spread, I felt obliged to spread a small spoonful of the jam on a bit of muffin. I was glad I did: the blackberry flavor was sweet and intense. "Utterly delicious!" I said. "It tastes like the berries have just now been picked!"

We bid a warm goodbye to the Blackwell family and drove off in our rental car — a tiny English Ford — for yet a bit more sight-seeing.

Wales is full of ancient castle ruins and we visited two: White Castle and Raglan Castle. They were complete with ducks swimming in a moat, and a heavy drawbridge leading to the stone castle itself. Although the interiors of these castles are crumbling ruins, enough remains of the sturdy exterior stone masonry and turreted towers to envision what they must have looked like in their heyday. My ten-year-old daughter had the time of her life running up the circular staircase to each tower; I could imagine her picturing herself a princess of eight-hundred years ago.

After a brief stop to stroll through the ruins of Tintern Abbey, we headed for the six-lane M-4 expressway.

It had taken me only a few hours in our little rented car to get used to the steering wheel on the wrong side; but somewhat longer getting used to driving on the wrong side of the road.

We had no strict itinerary. With only ten days of holiday our scope of travel was limited, so we concentrated on seeing the southern English countryside. We wanted to see the oldest cathedral in Britain — Winchester; and of course we wanted to see Stonehenge.

We saw Winchester Cathedral the second day and it was grand. Even ten-year-old Ann was impressed by this edifice dating from 1100. On succeeding days we ambled through cathedrals at Salisbury, Wells and Glastonbury, none of which seemed to compare with the magnificence of Winchester.

Stonehenge was well worth the visit. I guess no one will ever know how ancient people dragged twenty-ton blocks of stone all the way from Wales and then tipped them up on end to stand in a circle. The nearby village of Avesbury, with its thatched cottages, has a similar ring of stones, even more fascinating: the ring is much wider in diameter than Stonehenge; the ring of stones at Avesbury circles the entire village.

Aside from sight-seeing, what we enjoyed most was staying overnight at English farmhouse Bed and Breakfasts. Our very first night was at Folly Farm, owned by Mr. And Mrs. Kimber, near Crawley, Sussex.

The woman at the tourist information bureau in the village of Crawley gave us directions to the farm: "Go through the roundabout in Crawley, take the A272 and turn right about two-hundred yards past the Rock-and-Manger pub."

The Kimbers own several hundred hectares of land, milk one-hundred Friesian cows, and have a flock of two-hundred sheep. Mrs. Kimber was one of the few hostesses who also served an evening meal, at an extra charge of only one pound. She had roast pheasant.

Mr. Kimber was not around when we arrived for supper. "He is combining wheat," his wife explained. "He will be working in the field late tonight, until well after midnight. The weather has been terrible all through August—nothing but rain—so he has to take advantage of these few dry days we're having."

The farmers throughout Wiltshire and Hampshire were taking advantage of the good weather; everywhere we went we saw combines in wheat fields.

Next morning at breakfast, Mr. Kimber popped in for a brief visit. I chided him about burning the midnight oil.

"In more ways than one," he said, and laughed. "I'm burning oil in the grain dryer for the wheat too! That's our English weather for you!"

His views on the English dairy scene sounded familiar to me: "Our big problem is milk surplus."

A similar view was expressed by Mrs. Dyke of Townsend Farm at Croscombe, near the village of Shepton Mallet.

"The government is trying to urge farmers to get out of dairying," she said. "They are giving a bonus for each dairy cow shipped to slaughter. Farmers get around this by shipping a barren cow, then buying a springing heifer."

Another pleasant B & B visit was at the home of Cynthia and Colin Fletcher of Lower Foxhangers Farm in Wiltshire. A dozen years ago Colin worked as an exchange trainee on a Canadian farm for a period of eighteen months. He now raises beef cattle.

The thing we remember most about Foxhangers Farm was the evening stroll my daughter and I took along the nearby canal. (England is full of these narrow waterways, many of which are still used for barge traffic.)

Suddenly, three over-friendly goats emerged from the hedgerow that bordered the path along the canal. I picked up a hefty branch to discourage their advances. Our fears were not eased when an English family, also out for a stroll, met us and remarked, "Last week that goat pushed a woman into the canal!"

Also at Foxhangers we met another tourist, a white-haired English gentleman on holiday from his home in Yorkshire in the north. "It's the most beautiful part of England," he said. "All hills and dales and wild moors."

Since time was short, our visit to the moors was limited to

Exmoor at the southwest corner of Britain. Considered 'tame' compared to the Yorkshire moors; nevertheless we still thought it was ruggedly beautiful — a high, treeless land carpeted with purple heather on rolling hills. Ranging freely over the moor were flocks of sheep and wild ponies.

On a brief holiday, all the days seem to merge together. However, we still remember the ride in search of Lower Brown Farm. We drove over narrow English country roads that were bracketed on both sides with tall, clipped hedgerows; the road was so narrow, and the hedgerow so tall that we seemed to be driving through the tunnels of a maze.

Our directions were to drive through the hamlets of Wiveliscombe and Huish Champflower, but somehow or other, no matter how many turns I made, I kept ending up at the four corners of Wheddon Cross.

Finally, a young chap who was going our way led us on the right road. Mrs. Armitage, who presided at Lower Brown Farm provided a sumptuous evening meal topped off with plum pie hot from the oven.

After that feast, we went for our usual evening stroll. We watched the mist settle on the green hills, the Ayrshires grazing in the long pasture, the sheep baaing in the meadow. It was truly "a beauteous evening, calm and free." An English farming country evening.

LOCH LOMOND

"Put another log on the fire, Karen," said her mother. "It's getting chillier. This rain!"

Mrs. McLachlan buttoned up her cardigan and then went to open the door for her husband who was bringing in an armload of wood.

"Angus, you'll be that soaked!" she exclaimed, peering out into the darkness as he entered. "Goodness, it's bucketing out there!"

The fire soon was blazing again. Mrs. McLachlan poured the tea and passed around the tray of little cakes. With the long couch and two stuffed chairs drawn up before the fireplace, it was cozy in the parlor of the little cottage of Craighouse. The five of us — myself, my eleven-year-old daughter, and the three McLachlans — were quite comfortable.

We two Americans on holiday in Scotland were undaunted by the dismal weather. What if the relentless rain streamed down? It never could get through the thick shingles of Craig house, perched high above Loch Lomond.

Angus took the heavy metal poker and stirred the fire, shifting the slabs of peat and chunks of wood, encouraging the flames to blaze up higher. Then he leaned back and took a long, satisfying drink of hot tea.

"And did you know we have a monster here in Loch Lomond?" he inquired.

"No!" I answered in surprise. "I thought the monster was in Loch Ness, further north."

"Oh, there's been one seen here too." His eyes seemed to twinkle; was it merriment, or was it the firelight reflecting?

"Tell me about the Loch Ness monster," I asked. "Is it really some kind of deep water serpent or is it just a publicity stunt?"

"Oh, there have been many verified sightings," his wife was quick to reply.

"Last year some scientists even descended in a miniature submarine," Karen added.

"And what did they see?"

"Not much, I'm afraid. The lochs are very deep. And the water is murky at those depths."

Angus reached for another scone. "I don't believe they've ever found the bottom, actually. Those lochs could be as much as two miles deep, and according to one theory all the lochs are interconnected at a great depth."

"And what about you?" I asked. "Do you believe the monster is real?"

Angus shifted his position in the armchair, and this time there was no doubt about the twinkle in his eye. "I dunno. It's curious though how there seem to be a new flurry of sightings of the monster whenever the tourist season gets a bit slow."

I smiled. "Then you do think it's just publicity."

"I didn't say that. Maybe it's just coincidence."

These Scots! Their humor was so subtle you never knew when they were serious or when they were pulling your leg, or if they were merely being philosophical.

The dancing flames in the fireplace were mesmerizing. As I gazed at them, quite unexpectedly, a memory came flashing back to me of something that occurred many years ago at my farm.

It was a frosty December morning and my English pointer bitch had broken out of her kennel and gotten bred by the neighbor's Weimeraner. I had planned to mate her with another registered pointer; now my plans had all gone awry.

What to do? I asked our vet if there was a hormone injection he could administer to abort the mating, but at that time there was none.

"You will just have to let nature take its course," he said.

I can still see him looking at my pointer bitch with a quizzical smile and saying, "The best laid plans of mice and men..."

Doc Cairns was of Scottish descent, so it was natural for him to quote Robert Burns. And now here we were in the heart of Robert Burns country: Ayrshire and Dunbartonshire and the middle farming region that lies between the southern uplands and the northern highlands. I had wanted to see the countryside that produced one of the finest English lyric poets of the eighteenth century. And so I planned this trip to the British Isles. We were motoring in a rented car, and staying at farm bed-and-breakfasts all the way from the airport at Gatwick to the Scottish border. At long last, here we were, arrived at 'the bonnie bonnie banks of Loch Lomond.'

Robert Burns was the son of a poverty-stricken Scottish farmer who could afford to give the boy only the barest possible education. It looked like Robert would follow in the foot steps of oppressive poverty and mindless work.

But there were yearnings in him to reach out to something higher. To pass the monotonous time of day while he followed the horse and plow, Robert Burns sang to himself and composed bits of verse. Then by chance he read a book by a popular Scots writer who had begun to write in Scots dialect.

Robert recognized the possibilities at once and began jotting down his own verse in the dialect of an ordinary Scots farmer. A collection of his poems was printed in a small volume and became a best seller. It enabled him to pay off his father's debts, including the farm that had been mortgaged for many years.

One of Robert Burns' best known poems is "To a Mouse" and the subtitle is "On turning her up in her nest with the plow, November 1785." It shows the keen perception of a farm boy and his empathy with the humblest of nature's creatures, and also the ability to form a philosophical concept from so ordinary an incident as plowing up a field mouse nest.

It was from this poem my vet had quoted: "The best laid plans o' mice an' men, Gang aft agley."

And John Steinbeck borrowed it for the title of his heart-breaking play, "Of Mice and Men."

The last stanza of the poem is what I like the best, when Robert Burns tells the mouse:

> Still, thou art blessed compared wi' me!
> The present only toucheth thee.
> But och! I backward cast mine e'e,
> On prospects drear!
> And forward, though I canna see,
> I guess an' fear!

My reverie was interrupted by Angus who stirred the fire with the poker, sending a shower of sparks up the chimney.

"Let's hope this rain doesn't turn to snow," he said. "This late in the fall there always is that possibility."

I wondered what the lake would look like surrounded by snow-covered hills.

"Sure, it's a beautiful sight, the first snowfall," said Mrs. MacLachlan. "I remember one winter at 'Hug m' Knee' when the drifts were so deep we had to leave the car and walk."

"Hug my Knee?" I asked. "What's that?"

"That's when we go round before the New Year, wishing everyone well and having a drink with them."

"And we remember the long ago times," added Angus. "Aye, we remember the long ago times."

Old long ago. Or, as Robert Burns wrote, "Auld land syne." Has there ever been a song that has so thoroughly captured the hearts of men worldwide?

Should auld acquaintance be forgot,
And never brought to min'?
Should auld acquaintance be forgot,
And auld lang syne?

For auld lang syne, my dear,
For auld lang syne,
We'll tak a cup o' kindness yet
For auld lang syne.

Sung by men everywhere as the old year passes, it was first sung by a poor farm boy in Scotland as he followed the plow.

KINGS PARK FARM

The fog swirled over the top of the hill, making ghostly shapes of trees in the woodlot where I was loading a wagon with split chunks for the furnace. Thunder rumbled from the murky sky, an eerie sound even during a January thaw. The misty vapors dissolved and merged and dissolved again, giving me a momentary glimpse of the farm buildings far below in our little valley.

In an instant I was transported thousands of miles across the sea to another foggy view — a damp and clammy September morning in Scotland when my daughter and I leaned our elbows on the ramparts of Stirling Castle and peered over the cliff edge down at Stirling Plain, obscured by the fog. It was like a scene out of Macbeth. We half expected to hear the clank of swords, or see the three witches stirring their cauldron over a fire in the mist.

The mist we saw finally dissolved in the benevolent, though tardy, morning sunshine. Our patience was rewarded by a glorious view of King's Park Farm on the plain far below us. My daughter was ready with her camera to record the scene.

It was at King's Park Farm that we had spent the previous night at an English farm Bed and Breakfast. There had been other tourists spending the night at the farm: three English couples — farmers from the south of England on holiday. We were the only Americans.

We had a grand visit with them all over late evening tea and cakes provided by our hostess, Janet Johnston. Her husband, Robert, a tall burly Scotsman, kept the conversation flowing until well nigh midnight.

He had changed his farm work clothes for a clean shirt, flannel slacks and slippers. He settled his lanky frame into the comfortable sofa as he joined us tourists in the living room. In his deep voice he explained how the farmers brought the sheep down from the highlands for the winter.

"Aye. Two good dogs and a pair o' binoculars are worth three or four shepherds!"

"Do you have any preference as to the kind of dog you use to bring the sheep down?" asked Duncan, the farmer from Sussex.

"Aye." The r's rolled off Rob's tongue in broad Scottish accents. "A dark colored dog is preferred. That way the sheep make no mistake the animal they're seeing is a dog. You ken they 'ave been roaming free since spring and 'ave seen no other creatures but sheep since then."

"My wife and I wondered as we motored through the high lands how the sheep ever were brought down," Duncan continued. "Rugged country that, high enough to be called mountains, and covered thick with heather and gorse. Amazing how you ever find all the sheep."

Rob chuckled and took a deep drag on his cigarette. "It's no' an easy task when the sheep 'ave been roaming wild all summer."

There were eight of us tourists relaxing in the living room, and since we all happened to be farmers on holiday, conversation came easy. The others compared farming practices in the south of England with those in Scotland, as I listened intently.

Robert answered a battery of questions about soil types, plowing techniques and haymaking. Then he looked over at me and said, "Unless I miss my guess, we have another farmer here."

"A dairy farmer from America," I acknowledged. "In New York State."

Rob's face creased in a smile. "I thought so. I recognized the hands and arms of a farmer."

Now a flood of questions was directed to me by the others about dairy technology and crop production in the states.

Then I turned my attention once again to our host. "I suppose fall and winter are slack times of the year for you. Do you get a chance to travel then?"

"A wee bit. That is the time o' year our son and daughter concentrate on their dancing."

"They've done admirably well, judging by all these trophies

and medals," said Duncan's wife, indicating the glass-fronted cabinets that lined two walls of the living room.

There were literally hundreds of awards displayed behind the glass doors. I had noticed them when I first came in the room, but thought they were trophies for shooting, or sports.

"Aye. We're quite proud o' both Willie and Fiona. Willie has been dancing for nine years, and his sister for half as long."

I looked again at the cabinets and then noticed on the wall a color photograph of a teenage boy in kilts.

"That was taken two years ago when Willie won a dancing contest at the Highland Games," Rob said.

"What kind of dancing does he specialize in?"

"He started out in tap, then trained in traditional Highland dancing. But he also learned Scots country dancing, Russian Cossack, and Irish dances."

I mentioned that dancing seemed an unusual avocation for a farm family. Not so, Robert assured us.

"Music and dancing fit hand-in-glove with farming; the perfect relaxation from the physical labor and stress o' agriculture."

Rob ran a big callused hand through his hair and his thick black eyebrows met as his forehead furrowed in thoughtful recollection. "Music was the salvation o' me years ago when I came out o' the service and began working as a hired hand. Many was the time after riding the tractor all morning I was such a bundle o' nerves I could no' eat my lunch. I would go off by myself with the accordion and after playing for ten or fifteen minutes I could feel myself unwind."

"Do you play the bagpipes too?" Duncan asked.

"Aye, some. But they're a wee bit more difficult to master."

"How did your son happen to develop an interest in dancing?" asked James, the farmer from Essex.

"You might say it was by accident. As a wee lad he had a problem with his feet toeing in some. Our doctor suggested dancing lessons would help correct the problem. In desperation, when he was twelve years old he took up tap dancing."

"And did it help?"

"Aye. And as a bonus you might say, Willie discovered he had a real talent for dancing."

"Admirable!"

"Within five years he became accomplished enough to be invited on tours overseas. He's been once with a dance group to France and Belgium, and spent a month touring the East Coast of America."

"In addition now for the past two years he has been on a weekly TV variety show on the BBC with a dance group. The name of the show is Thing-a-Majig and it is similar to the TV show in America called Hootnanny," he said, turning to me.

"That sounds like a pretty demanding schedule," I said. "How does he find time to do all that and help on the farm?"

"There's not much time in the summer o' course with haying and crops. The rehearsals for the TV show require a considerable amount o' driving, but we don't mind that, and fortunately the shows are taped in the fall when farm work is slack."

"It helps to have a supportive mother and father," Duncan said.

"We're proud to help. And we're proud that in addition, Willie attended Falkirk Technical College and received a degree in agriculture this year."

Rob put out his cigarette. "If you can suffer more o' a father's pride for awhile, I've got some video tapes of Willie's TV program. I think you will enjoy seeing them."

"By all means!"

We sat back and enjoyed bits of half a dozen shows. It was while we were having late tea, prior to retiring for the night, that Willie poked his head in the doorway. He was a tall and slender young man of twenty-one.

"Father, the animals are all fed," he said. "And the dogs are taken care of too."

We chatted with Willie for a few minutes and learned that he is booked to join another dance tour group to America.

All in all it was a very enjoyable stay with the Johnston family. To make it even more memorable, the following morning we were awakened at eight o'clock with music to remind the sleepy tourists that breakfast was ready. The music was the skirl of bagpipes.

We were effusive in our compliments for the hearty breakfast. Rob and his wife acknowledged our thanks. Rob said, "I tell everyone that's how we make our living — bed and breakfast for the animals, and bed and breakfast for the tourists!"

MONGOLIAN ADVENTURE

Mongolia. The name conjures up visions of Genghis Khan's fierce cavalry, the Golden Horde, that conquered half the world in the 13th century. It stirs ancient memories of fabulous Xanadu, the exotic court built by his grandson, Kublai Khan, immortalized in poetry by Coleridge. It was to Xanadu that Marco Polo journeyed with his uncle's caravan, bringing back tales of its splendor, along with silks and spices. I cannot believe I am here. It seems that I have not only stepped into another country, but another century.

Lifelong habit as a dairyman has awakened me early. I zip

open my tent and look outside. The full moon setting over blue hills in the west illuminates the scene: the pearly gray domes of three gers, the circular tents known as yurts in other parts of Asia, and on the grassy plain nearby are slumbering herds of cattle, sheep and goats. The low door of a ger opens and Mr. Nyamkhuu comes out into the predawn light to survey his domain.

Mr. Nyamkhuu is a livestock breeder of Lun *sum* (district) in a central *aimag* (province) of Mongolia. He owns 60 cows, 200 sheep, 100 goats, 120 horses and 12 camels. His extended family of two married sons with their wives and children help him manage these herds. Theirs is a nomadic life, following their animals as the grass grows. Half of the 2.2 million Mongolians are herders, living a pastoral existence.

We three Americans came here yesterday, riding in a Russian jeep across high plains stretching endlessly westward from the capital, Ulaan Baatar. With an average elevation of 5,200 feet, Mongolia is one of the highest countries in the world . Much of it is steppe — a high grassland prairie. Trees cover less than 10 percent of the land, mostly in the north. Along the western border are the Altai mountains, permanently snow-covered.

We passed huge fields of wheat, reminders of the collectivefarms that appeared briefly during the years of Soviet influence, only to disintegrate when the USSR bubble burst in1991. The abundance of bird life was astounding: storks and sandhill cranes near every marsh, eagles and hawks on the low bushes or rock outcroppings, preying on the prairie's profusion of mice and marmots. Mongolia is a naturalist's paradise, mainly because industry and its pollution are almost nonexistent.

Occasional flocks of sheep or herds of cattle seemed like ants in the distance, and the white gers of their herders looked like mushrooms, accentuating the vastness of the panorama.

Idevkhten, our driver, headed for one of these clusters of gers so that we could get a close-up look at a herder's life. When we arrived, Mr. Nyamkhuu gave Idevkhten the traditional male greeting, *zolgob*—something like shaking hands. The two men approached each other, palms of their hands facing up, and grabbed each other's elbows for a moment. The apparent origin of this greeting was to show that neither man was carrying a weapon up his sleeve.

We were invited into the ger and given a refreshing drink from a bowl of *airag*— fermented mare's milk (koumis in Russian). Orgilmaa, our translator, fielded the polite questions Mr. Nyamkhuu directed at us through her. "How is the weather in your country?" "Did you winter well?"

It's not surprising weather is a prime topic of conversation, considering that Mongolia has one of the harshest climates in the world. Winters are fierce, with temperatures dipping to minus 50 degrees fahrenheit and bitter winds blowing in from the Siberian steppes. The brief summers are pleasant, with highs near 70 degrees, although evenings are chilly because of the high altitude. The rainy season is July to September, but showers tend to be brief, and Mongolia is, on the whole, an arid country.

Mr. Nyamkhuu's ger was a spacious felt tent over a latticework frame and a wood floor covered with rugs. A ger is very practical for a family on the move. The whole thing can be dismantled in a few hours, loaded on a wagon or truck, and moved to a new location when the livestock must be moved to fresh grass. Mr. Nyamkhuu makes the move about once a month. All pasture land in Mongolia is state-owned, or common land. A herder may move his livestock anywhere within the boundaries of his sum. For this privilege he pays a livestock tax of 50 tugrik per year for each cow, or group of nine sheep or goats.

Since it was late in the afternoon, Mr. Nyamkhuu invited us to pitch our tents nearby and spend the night. We were invited into his ger for an evening meal of mutton soup with noodles, and salty milk tea, cooked on an iron stove in the center of the ger.

In the morning we shared his familys' breakfast of milk tea and biscuits spread with *urum,* or white butter. Urum is made by boiling milk and skimming off the thick skin on top. Another milk product is *aaruul.* This is milk curds, dried in the open air and sun. Dried aaruul curds are as hard as marbles and have to be sucked like candy,

Breakfast over, Mr. Nyamkhuu's two sons saddled their horses and rode off to bring in the main herd, which had been grazing during the night in the foothills a mile away across the plains. His wife and daughters-in-law got stools and pails and proceeded to milk a few of the cows. This accomplished, they carried two 20-liter cans of milk to the side of the wagon road. Soon a tanker appeared, pulled by a Belarus tractor. The cans were hoisted up, the milk dumped in the tanker, destined for the city. The price paid by the state is pitifully low. Indeed, most rural Mongolians live outside the cash economy, depending on barter or the black market.

A low rumble of thunder announced the arrival of the horse herd. Mares, foals, stallions and geldings splashed into the shallow water of a nearby lake and drank their fill. Then the two men began sorting out the milking mares from the foals. Each man used a uurik— a long, limber wooden pole with a noose at the end, galloping after an animal and snaring it. The women milked the mares, adding fresh milk to the airag already fermenting in the wooden barrel.

As might be expected, rural Mongolian families eat mostly meat and dairy products. In winter, meat dominates the diet, while in summer they prefer dairy products. Although in the cities inflation is rampant and food is scarce, rural Mongolians survive as they have for centuries, on the land that nourishes their livestock.

ALASKA BUSH PILOT

From Alaska bush pilot to Virginia dairyman is quite a change. Ben Hershberger spent sixteen months in Alaska as a bush pilot around 1981. When he read my June 1984 Dairyman's Journal article about the 'Musk Ox Producers' Cooperative', he wrote me an interesting letter telling about the time he flew some Alaska Fish and Game Department personnel over Nelson island for a musk ox count during the winter of 1981.

He wrote: "Nelson Island lies between Nunivak Island and Alaska's Bering Sea coast. Around 1970 a herd of ten musk ox from Greenland was transplanted to Nelson Island. Over the next ten years the transplanted herd grew from ten head to over two hundred-fifty head.

"Nelson Island is just inland from Nunivak Island. It hardly seems like an island because it is separated from the mainland only by a river. The villages of Tununak, Toksook and Nightmute are located on Nelson Island. Our family of five lived in a trailer court at Bethel for about a year — quite an experience!"

Ben's mention of Bethel conjures up a vision of this little village (population 3,681)on the banks of the mighty Kuskokwim River that empties into the Bering Sea adjacent to the equally mighty Yukon River. Together these two rivers spread out and create a mammoth delta, a gigantic mud pie that makes the Mississippi delta look like a frog pond.

The Yukon-Kuskokwim delta is dotted with myriad swamps and lakes teeming with fish, marine mammals and birds. This means plenty of food and skins for the Yup'ik Eskimos who live in countless tiny villages throughout the delta region. Roads are non-existent there. Travel over the entire region is either by boat or small plane. So I could understand thoroughly when Ben continued:

"We did charter flying. Since it is the only way to go, there was a lot of it to do. During the summer months we were very busy, with daylight lasting until around eleven o'clock at night. However, we were not far enough north to see the midnight sun. My flying in Alaska mostly was Cessna 207's, 210's, 185's and a Seneca II."

In 1983 Ben and his family returned to his wife's family's dairy farm in Virginia.

"Our oldest son is eleven, and another is nine. So already we have some help for the dairy farm. We are operating only a small forty-acre farm at present, but have an additional two hundred acres rented. We started milking in April and now are milking thirty cows, with plans for forty.

"I wouldn't mind hearing more of your Alaska experience, and I would like to see an article on the demise of dairying in Alaska."

I had mentioned in an earlier Dairyman's Journal article that the dairy industry in Alaska declined between 1970 and 1980 and almost expired. Here is a brief history.

Dairy farming really began in earnest in Alaska with the founding of the Matanuska Colony in 1935. As you may recall, the colony consisted of two hundred Midwestern families, escaping from the ravages of the Dust Bowl.

Each family was given a homestead of forty acres of uncleared land. Many of the families prospered and diversified into all aspects of farming: vegetables, livestock and dairy.

By 1960 there were fifty-two dairy farms in the Matanuska Sustina valley. Yet by 1980 their number had declined to only ten dairy farms, with only about twelve-hundred cows left. With milk selling in Alaska stores for ninety-nine cents a quart you would think the dairy industry would prosper. Why then the decline?

Principal villain was one most farmers are familiar with: urban sprawl. The villages of Palmer and Wasilla, formerly quiet rural towns in the Matanuska-Sustina valleys, are fast becoming bedroom suburbs of mushrooming Anchorage, which lies only forty commuter miles to the south. Tract-after-tract of houses are being constructed in the birch and spruce woods of the area.

Many farms were lost to this type of development, and to

speculators. As established dairymen retired or died, their farm land was snapped up by the real estate industry which priced it out of the reach of other farmers.

Even today, with the Point MacKenzie dairy project well into its third year — and located only a few miles across the water of Knik Arm from Anchorage—the controversy rages. There is talk of building a bridge across Knik Arm. A feature story in the newspaper recently concerned a borough meeting to discuss a proposed agricultural district for the Palmer and Wasilla areas. The meeting brought heated debate. Those who are against zoning for ag use say that if the state wants to restrict the land, it should buy the ag rights to that land.

Another reason for the decline of dairy farming was the pressure of 'unfair' competition of milk from Seattle (shipped to Alaska by boat from dairy farms in Washington and Oregon). Supermarkets had to buy the Seattle milk outright; if they did not sell it before it turned sour, they were stuck with it. On the other hand milk from the two Anchorage processing plants — North Star Dairy and Matanuska Maid — was taken on consignment; any milk not sold by expiration date was taken back by North Star and Matanuska Maid.

Naturally, the supermarkets pulled the Seattle milk up to the front of the dairy case and pushed Alaska milk to the rear, so you can imagine what that did to the sale of Alaska milk.

Things are changing though. Alaskans have proven that they prefer Alaska-produced milk. Heretofore there have been no state regulations on expiration dates for bottled milk. With Seattle milk taking upwards of eight days' shipping by boat, this naturally worked to Seattle's advantage. When expiration date-stamping for bottled milk becomes mandatory, Seattle milk that is eight days old will have a more difficult time competing in the dairy case with fresher Alaska milk.

I cannot conclude this article without mentioning the third milk processing plant located further north in Alaska at Delta Junction. Northern Lights Dairy is owned and operated by the Don Littleman family. Don buys milk from the only two dairy farmers in the Delta Junction area: Neil Schenk and Wayne Crowson. He bottles the milk and sells some of it locally in Delta Junction (population 1,044); the balance is shipped one hundred miles further north to Fairbanks.

Don is looking for more milk to supply his expanding market. He could take on another dairy or two. So...if you are looking for adventure.... Head for the land of the Northern Lights!

POINT MACKENZIE

"So you're a journalist and you have just arrived here at Point MacKenzie?" John was backing his car out of the garage when I pulled in his yard. "My wife and I are about to leave for Anchorage on business this morning, but we can spare a few minutes."

He led the way into the kitchen where his wife was busy with last-minute things. "Mary, this gentleman is visiting the new farms to get first-hand information on the dairy project."

"Well, I hope he can correct some of the overly optimistic articles about dairying in Alaska that we've read lately in the farm magazines — especially the "Dairyman's Journal" articles by that Richard Triumpho. He painted far too rosy a picture!"

"I guess I should plead guilty," I said, feeling somewhat flustered at the frank remark, and then introduced myself. "I am Richard Triumpho. My reason for coming to Alaska is to see with my own eyes what progress is being made here on the Point MacKenzie project."

Thereupon, we all sat down at the kitchen table and began a lively conversation.

"To begin with, I don't even know if we are going to be milking cows in that thing." John gestured out the kitchen window at the brand new free-stall barn. "Our cows are due to arrive at the end of September, and we don't even know if we are going to have a market for our milk!"

"Why?" I asked.

"Because Matanuska-Maid of Anchorage, the sole processing cooperative here, has filed for Chapter Eleven bankruptcy. In order to help the co-op reorganize and survive the dairymen have been asked to take a cut in the price we receive for our milk— a cut of from two dollars to four dollars a hundredweight."

"What price are dairymen receiving now?"

"Twenty-two dollars and seventy cents a hundred-weight."

So the cooperative was in trouble. It seemed like an old familiar song to a dairyman like myself from the 'lower 48': co-op leadership over the years in the hands of the few, not enough questions asked by the members, resulting in a weakening equity position, and so on.

John and Mary moved up from Minnesota a year ago to manage this new dairy farm for an Anchorage investor who had

Fields Afar

bought the land in the 1982 Alaska lottery. Everything back then looked rosy: the Alaska state government had millions of dollars of surplus money from North Slope oil; and the Alaska Ag Council was given the green light to develop the Point MacKenzie dairy project. This ambitious scheme to make the state self-sufficient in milk production was carving twenty thousand acres of farmland out of the virgin spruce forest.

This farm, like the others in the project, consisted of a full section — six hundred and forty acres — and most of it already was cleared of trees. John had two hundred acres of wheat and barley planted between the windrowed 'berms' of trees and roots. The free-stall barn construction was nearly complete, needing only finishing touches in the milking parlor. And now this new problem with the co-op left everything up in the air.

In concluding our discussion around the table, John was cautiously optimistic. "I don't see how the state can allow Mat-Maid to go under, especially with all the money the Alaska Revolving Loan Fund now has invested in the Point MacKenzie project."

Taking my leave of John and Mary, I continued my drive down the wide gravel road bisecting this wilderness project. I passed other farm parcels, which were in various stages of land clearing and building construction. The whole area had the air of a frontier. Yes, Alaska still was 'The Last Frontier'.

On both sides of the road the spruce forest had been pulled down, and bulldozers had pushed the tangled mess of trees and roots into long windrowed berms which stretched to the horizon. Between these berms, the field strips of cleared land — each strip was one hundred-fifty foot wide and anywhere from one mile to two miles long — were growing healthy crops of wheat, oats and barley. About every quarter mile, strips of standing trees had been left to serve as windbreaks.

"I timed myself, and it took fifty-six minutes for me to make one trip around the field with the four-wheel-drive tractor pulling the heavy ground-breaking disks," said Tom Rogers, the next dairyman I visited.

The field Tom was disking was one hundred-fifty foot wide and two miles long, and straight as an arrow. His farm parcel, a full section, measures one-half mile wide by two miles long.

Tom and his wife retired a couple of years ago from their farm in the upper Midwest, but came out of retirement to run this new farm for his brother-in-law who is a restaurateur in Anchorage.

The Rogers farm is the second farm to actually be shipping milk from the Point MacKenzie project. Last fall, as soon as their one hundred cow free-stall barn was complete, Tom bought a herd of forty-eight cows from a farmer near the city of Palmer. Before long he plans to be operating at capacity.

When I came in the yard, Tom had just driven through the barn with the feed wagon, delivering oat silage he had put up in one of two long plastic 'baggers'. The herd also gets high moisture barley, which Tom buys from the Delta project further north.

Commenting on some of the problems he encountered in building a Grade A dairy out of the wilderness in just two years Tom mentioned moving soil.

"The topsoil here is about twenty inches deep — a peaty humus type typical of a cleared forest situation — and under that is gravel, three hundred feet deep. Consequently, when we excavated to set poles for the barn, we ended up moving far more soil than we intended because the gravel kept caving in.

However, this same gravel means the soil is well drained. And good water is only fifty feet below the surface."

Further down the road from the Rogers place, I stopped at a clearing where a Quonset-type free-stall barn was being built. The owner wasn't there, but a tall strapping young fellow left his job of tightening bolts to talk to me. Bill Green was his name; he had moved up from Iowa six months ago to become herdsman-manager here. When he heard my name, he grabbed my hand and gave it a hearty shake.

"Are you Richard? You're the reason I'm here! I read your articles on Alaska last spring and decided this was the place for me!"

Bill said he had been farming for a dozen years in Iowa. "I invested in the farm and equipment in 1970 when real estate was sky-high and the economy was booming. Then you know what happened: interest rates climbed in the late 70's and real estate dropped in value. The last few years we were paying unbelievable amounts of interest on the mortgage. When I got up in the morning, even before I put my socks on, I knew I had to make a hundred dollars that day before I could start working for myself. There was no future in that. That's why I'm here."

My last stop that day was at six o'clock. Fred and Rachel were getting ready for their supper at a picnic table outside their camper; the camper had been their home for the six weeks they had been here. The biggest bulldozer I ever saw was grinding to a halt in the clearing across the road.

"He's been pushing up trees," Fred commented. "He can really make things move in one day."

Fred and Rachel had farmed in Iowa; they both also were schoolteachers, and they came to Alaska looking for teaching jobs. They stopped at a real estate office in Wasilla to inquire about housing. It turned out that the realtor had bought this farm parcel in the Alaska State lottery and didn't know a thing about dairy farming. So, after talking with Fred and Rachel, he hired Fred as manager.

"We got here in June," Fred told me. "Had six weeks to prepare a farm plan, complete with building costs to the last nut and bolt, and a clearing and planting schedule; this all was submitted to the Revolving Loan Fund for approval. We hope to get the okay from them in time to build the barn before freeze-up; otherwise we'll never make our schedule next year."

I asked about the progress he made in land clearing.

"Great," he said. "That ship's anchor chain they use to knock down trees really is something. Have you seen it?"

I told him I hadn't seen one, but I had heard how they used it: A long chain was stretched between two bulldozers, which dragged it though the forest; the trees, being shallow-rooted, toppled like mown wheat.

"There's a chain lying alongside the road a mile or so from here," Fred told me. "You should stop and heft it. You can barely lift the links on one end. The chain is two hundred forty feet long and weighs seven tons."

When I asked Fred what he thought of the problems with the Mat-Maid Cooperative, he did not think they were insurmountable. His whole mood was one of optimism.

"One thing is sure," he said. "I'll never see anything like this again in my lifetime. From wilderness to Grade A dairy in less than two years — a real adventure."

Listening to him, I was reminded of what Helen Keller said: Life is either a great adventure or it is nothing at all.